The Water Legacies of Conventional Mining

The impact of mining is too big to ignore in a world of oversubscribed water. This is true of conventional mining as much as – or even more than – hydraulic fracturing (fracking). The legacy issues of such mining on water have not been fully appreciated, especially the irretrievable effects mining has had on communities and ecosystems around the world through its impact on water. Yet this is not an 'us-or-them' problem: the wealth, influence and technical knowledge of mining interests can and must be part of the solution. All of the contributions to this volume either consider the deficiencies of existing governance structures and the need for better ones or explore the use of new techniques to identify and evaluate social and environmental impacts.

The chapters in this book were originally published in the journal *Water International*.

James E. Nickum is an institutional economist, affiliated with the Centre for Water and Development at SOAS, University of London, UK, and the University of Hong Kong, Hong Kong, and is Editor-in-Chief of *Water International*.

David B. Brooks is a natural resource economist working mainly in North America and the Middle East; he is affiliated with the International Institute for Sustainable Development in Manitoba, Canada.

Anthony Turton is a water strategist in both the academic and commercial world, affiliated with the Centre for Environmental Management at the University of Free State, Bloemfontein, South Africa.

Surina Esterhuyse is a geohydrologist working on oil and gas extraction in South Africa, and affiliated with the Centre for Environmental Management at the University of Free State, Bloemfontein, South Africa.

Routledge Special Issues on Water Policy and Governance

https://www.routledge.com/series/WATER

Edited by
Cecilia Tortajada *(IJWRD), Third World Centre for Water Management, Mexico*
James E. Nickum *(WI), International Water Resources Association, France*

Most of the world's water problems, and their solutions, are directly related to policies and governance, both specific to water and in general. Two of the world's leading journals in this area, the *International Journal of Water Resources Development* and *Water International* (the official journal of the International Water Resources Association), contribute to this special issues series, aimed at disseminating new knowledge on the policy and governance of water resources to a very broad and diverse readership all over the world. The series should be of direct interest to all policy makers, professionals and lay readers concerned with obtaining the latest perspectives on addressing the world's many water issues.

Water Infrastructure
Edited by Cecilia Tortajada and Asit K. Biswas

Frontiers of Land and Water Governance in Urban Regions
Edited by Thomas Hartmann and Tejo Spit

The Water-Energy-Food Nexus in the Middle East and North Africa
Edited by Martin Keulertz and Eckart Woertz

Sustainability in the Water-Energy-Food Nexus
Edited by Anik Bhaduri, Claudia Ringler, Ines Dombrowsky, Rabi Mohtar and Waltina Scheumann

Transboundary Water Cooperation
Principles, practice and prospects for China and its neighbours
Edited by Patricia Wouters, Huiping and James E. Nickum

Water Reuse Policies for Potable Use
Edited by Cecilia Tortajada and Choon Nam Ong

Hydrosocial Territories and Water Equity
Theory, governance and sites of struggles
Edited by Rutgerd Boelens, Ben Crow, Jaime Hoogesteger, Flora E Lu, Erik Swynedouw and Jeroen Vos

Energy for Water
Regional case studies
Edited by Christopher Napoli

The Water Legacies of Conventional Mining

Edited by
James E. Nickum, David B. Brooks, Anthony Turton and Surina Esterhuyse

Routledge
Taylor & Francis Group

LONDON AND NEW YORK

First published 2017 by Routledge

2 Park Square, Milton Park, Abingdon, Oxfordshire OX14 4RN
52 Vanderbilt Avenue, New York, NY 10017

Routledge is an imprint of the Taylor & Francis Group, an informa business

First issued in paperback 2018

British Library Cataloguing in Publication Data
A catalogue record for this book is available from the British Library

ISBN 13: 978-1-138-28871-3 (hbk)
ISBN 13: 978-0-367-22046-4 (pbk)

Typeset in TimesNewRomanPS
by diacriTech, Chennai

Publisher's Note
The publisher accepts responsibility for any inconsistencies that may have arisen during the conversion of this book from journal articles to book chapters, namely the possible inclusion of journal terminology.

Disclaimer
Every effort has been made to contact copyright holders for their permission to reprint material in this book. The publishers would be grateful to hear from any copyright holder who is not here acknowledged and will undertake to rectify any errors or omissions in future editions of this book.

Contents

Citation Information   vii

Notes on Contributors   ix

1 Water legacies of conventional mining: introduction   1
   David B. Brooks, James E. Nickum, Anthony Turton and Surina Esterhuyse

2 Untying the Gordian Knot: unintended consequences of water policy
   for the gold mining industry in South Africa   7
   Anthony Turton

3 Mine site water-reporting practices, groundwater take and governance
   frameworks in the Hunter Valley coalfield, Australia   28
   Wendy Timms and Cameron Holley

4 Non-discrimination and liability for transboundary acid mine drainage
   pollution of South Africa's rivers: could the UN Watercourses Convention
   open Pandora's mine?   48
   Rémy Kinna

5 A pilot study of the Social Water Assessment Protocol in a mining
   region of Ghana   69
   Anastasia N. Danoucaras, Alidu Babatu Adam, Kathryn Sturman,
   Nina K. Collins and Alan Woodley

6 Unconventional oil and gas extraction in South Africa: water linkages
   within the population–environment–development nexus and its policy
   implications   86
   Surina Esterhuyse, Nola Redelinghuys and Marthie Kemp

7 Lessons from Yanacocha: assessing mining impacts on hydrological
   systems and water distribution in the Cajamarca region, Peru   103
   Diana Vela-Almeida, Froukje Kuijk, Guido Wyseure and Nicolas Kosoy

CONTENTS

8 Disputes over land and water rights in gold mining: the case of
   Cerro de San Pedro, Mexico   124
   Didi Stoltenborg and Rutgerd Boelens

9 Mining and campesino engagement: an opportunity for integrated
   water resources management in Ancash, Peru   145
   Robert Patrick and Lalita Bharadwaj

10 Questioning the effectiveness of planned conflict resolution strategies
   in water disputes between rural communities and mining companies
   in Peru   160
   Milagros Sosa and Margreet Zwarteveen

11 Predicting water quality associated with land cover change in the
   Grootdraai Dam catchment, South Africa   178
   Anja du Plessis, Tertius Harmse and Fethi Ahmed

12 Assessing the existing knowledge base and opinions of decision
   makers on the regulation and monitoring of unconventional gas
   mining in South Africa   195
   Surina Esterhuyse, Marthie Kemp and Nola Redelinghuys

13 A water-energy-food security analysis tool for mining in Suriname:
   operationalizing the Mining Policy Framework of the Intergovernmental
   Forum on Mining, Minerals, Metals and Sustainable Development   209
   *Dimple Roy, Darren Swanson, Carter Borden, Alec Crawford,
   Livia Bizikova and Gabriel Huppe*

   Index   219

Citation Information

The chapters in this book were originally published in various editions of *Water International*. When citing this material, please use the original page numbering for each article, as follows:

Chapter 2
Untying the Gordian Knot: unintended consequences of water policy for the gold mining industry in South Africa
Anthony Turton
Water International, volume 41, issue 3 (2016) pp. 330–350

Chapter 3
Mine site water-reporting practices, groundwater take and governance frameworks in the Hunter Valley coalfield, Australia
Wendy Timms and Cameron Holley
Water International, volume 41, issue 3 (2016) pp. 351–370

Chapter 4
Non-discrimination and liability for transboundary acid mine drainage pollution of South Africa's rivers: could the UN Watercourses Convention open Pandora's mine?
Rémy Kinna
Water International, volume 41, issue 3 (2016) pp. 371–391

Chapter 5
A pilot study of the Social Water Assessment Protocol in a mining region of Ghana
Anastasia N. Danoucaras, Alidu Babatu Adam, Kathryn Sturman, Nina K. Collins and Alan Woodley
Water International, volume 41, issue 3 (2016) pp. 392–408

Chapter 6
Unconventional oil and gas extraction in South Africa: water linkages within the population–environment–development nexus and its policy implications
Surina Esterhuyse, Nola Redelinghuys and Marthie Kemp
Water International, volume 41, issue 3 (2016) pp. 409–425

Chapter 7

Lessons from Yanacocha: assessing mining impacts on hydrological systems and water distribution in the Cajamarca region, Peru
Diana Vela-Almeida, Froukje Kuijk, Guido Wyseure and Nicolas Kosoy
Water International, volume 41, issue 3 (2016) pp. 426–446

Chapter 8

Disputes over land and water rights in gold mining: the case of Cerro de San Pedro, Mexico
Didi Stoltenborg and Rutgerd Boelens
Water International, volume 41, issue 3 (2016) pp. 447–467

Chapter 9

Mining and campesino engagement: an opportunity for integrated water resources management in Ancash, Peru
Robert Patrick and Lalita Bharadwaj
Water International, volume 41, issue 3 (2016) pp. 468–482

Chapter 10

Questioning the effectiveness of planned conflict resolution strategies in water disputes between rural communities and mining companies in Peru
Milagros Sosa and Margreet Zwarteveen
Water International, volume 41, issue 3 (2016) pp. 483–500

Chapter 11

Predicting water quality associated with land cover change in the Grootdraai Dam catchment, South Africa
Anja du Plessis, Tertius Harmse and Fethi Ahmed
Water International, volume 40, issue 4 (2015) pp. 647–663

Chapter 12

Assessing the existing knowledge base and opinions of decision makers on the regulation and monitoring of unconventional gas mining in South Africa
Surina Esterhuyse, Marthie Kemp and Nola Redelinghuys
Water International, volume 38, issue 6 (2013) pp. 687–700

Chapter 13

A water-energy-food security analysis tool for mining in Suriname: operationalizing the Mining Policy Framework of the Intergovernmental Forum on Mining, Minerals, Metals and Sustainable Development
Dimple Roy, Darren Swanson, Carter Borden, Alec Crawford, Livia Bizikova and Gabriel Huppe
Water International, volume 41, issue 7 (2016) pp. 1035–1043

For any permission-related enquiries please visit:
http://www.tandfonline.com/page/help/permissions

Notes on Contributors

Alidu Babatu Adam is a PhD student at the Sustainable Minerals Institute, University of Queensland, Brisbane, Australia.

Fethi Ahmed is Professor and Head of the School of Geography, Archaeology and Environmental Sciences at the University of the Witwatersrand, Johannisburg, South Africa.

Livia Bizikova is Director, Knowledge for Integrated Decisions at the International Institute for Sustainable Development (IISD), Winnipeg, Canada.

Lalita Bharadwaj is Associate Professor at the School of Public Health, University of Saskatchewan, Canada.

Rutgerd Boelens, Professor of 'Water Management and Social Justice' at the Water Resources Management Group in Wageningen University, the Netherlands, is also Special Professor in the Department of Human Geography, Planning and International Development (GPIO) at the University of Amsterdam. He holds the Extraordinary Chair in 'Political Ecology of Water in Latin America' in the interuniversity Centre for Latin American Research and Documentation (CEDLA) and is a Visiting Professor at the Catholic University of Peru.

Carter Borden is Vice President of Centered Consulting International LLC, Boise, Idaho, USA.

David B. Brooks is a natural resource economist working mainly in North America and the Middle East; he is affiliated with the International Institute for Sustainable Development in Manitoba, Canada.

Alec Crawford is Senior Researcher with IISD's Resilience and Economic Law and Policy Program, Winnipeg, Canada.

Nina K. Collins is a Research Analyst at the Sustainable Minerals Institute, University of Queensland, Brisbane, Australia.

Anastasia N. Danoucaras at time of writing was a Post-doctoral Research Fellow at the Sustainable Minerals Institute, University of Queensland, Brisbane, Australia.

Anja du Plessis is Senior Lecturer at the University of South Africa.

Surina Esterhuyse is a geohydrologist working on oil and gas extraction in South Africa, and is affiliated with the Centre for Environmental Management at the University of Free State, Bloemfontein, South Africa.

Tertius Harmse is Emeritus Professor of the Department of Geography, Environmental Management and Energy Studies, Kingsway Campus, University of Johannesburg, South Africa.

Cameron Holley is Associate Professor at the Connected Waters Initiative Research Centre, University of New South Wales, Australia.

Gabriel Huppe is an associate with IISD in Winnipeg, Canada, and is pursuing a PhD at the Ivey Business School at Western University, London, Canada.

Marthie Kemp is Senior Professional Officer at University of the Free State, Bloemfontein, South Africa.

Rémy Kinna is an Australian international water law, policy and governance specialist, currently based in Phnom Penh, Cambodia.

Nicolas Kosoy is Associate Professor at the Department of Natural Resource Sciences, McGill University, Québec, Canada.

Froukje Kuijk is Consultant in Integrated Water Resources Management and Assistant Programme Specialist of UNESCO-IHP projects in Santiago, Chile.

James E. Nickum is an institutional economist, affiliated with the Centre for Water and Development at SOAS, University of London, UK, and the University of Hong Kong.

Robert Patrick is an Associate Professor at the Department of Geography and Planning, University of Saskatchewan, Canada.

Nola Redelinghuys is a Research Associate at the Department of Sociology, University of the Free State, Bloemfontein, South Africa.

Dimple Roy is the Director of the Water Program at IISD, Winnipeg, Canada.

Milagros Sosa is a PhD student at the Water Resources Management Group, Wageningen University, the Netherlands.

Didi Stoltenborg is a Junior Researcher and Lecturer at the Soil Physics and Land Management Group, Wageningen University, the Netherlands.

Kathryn Sturman is a Senior Research Fellow at the Sustainable Minerals Institute, University of Queensland, Brisbane, Australia.

Darren Swanson is Director of Novel Futures Corporation, Winnipeg, Canada, and an associate with the IISD.

Wendy Timms is a hydrogeologist, environmental engineer and Director of Postgraduate Studies at the School of Mining Engineering, UNSW Australia.

Anthony Turton is a water strategist in both the academic and commercial world, affiliated with the Centre for Environmental Management at the University of Free State, Bloemfontein, South Africa.

Diana Vela-Almeida is a Researcher for the Ecological Economic Research group at McGill University, Québec, Canada.

Alan Woodley is a member of staff at the Sustainable Minerals Institute, University of Queensland, Brisbane, Australia.

Guido Wyseure is Associate Professor at the Soil and Water is Management Division, Katholieke Universiteit Leuven, Belgium.

Margreet Zwarteveen, an irrigation engineer and social scientist, is Professor of Water Governance in the Integrated Water Systems and Water Governance, a department of UNESCO-IHE, Delft, the Netherlands.

Water legacies of conventional mining: introduction

David B. Brooks, James E. Nickum, Anthony Turton and Surina Esterhuyse

In the past, the legacy of mining's effect on water use and disposal was not fully appreciated. Stakeholder involvement was primarily shareholder and government involvement. Local communities gained wider notice only as victims of a disaster such as the collapse of a mountain of mining waste (for example, Aberfan, Wales, in 1966 and Merriespruit, South Africa, in 1994). Today these impacts cannot be ignored. Many people inside and outside the minerals industry recognize that conventional mining does have a legacy that goes beyond the minerals and metals it has withdrawn from the earth. More often than not, that legacy reflects the way that the mining industry's use and disposal of water has irretrievably affected communities and ecosystems. Moreover, around the world, this "mining–water nexus" can be shown to have social and political dimensions. At the same time, we cannot ignore that the wealth, influence, and technical knowledge of mining interests can and must be part of the solution. For that reason, the editors of this book were after two sorts of studies:

(1) Those in which mining is clearly part of a problem that needs to be highlighted and addressed.
(2) Those in which mining is clearly part of a solution that is either being developed or already exists.

Those two foci reflected recognition that the interaction between mineral extraction and water resources does indeed pose problems and, more importantly, that the knowledge base and implementation capacity for finding sustainable solutions to these problems ***must arise largely from within the mining sector itself***. Although the mining sector traditionally externalized their costs to the environment and society, this sector is now increasingly involved in contributing to the development of future solutions to current water constraints in their areas of operation. Finding a balance between solving mining-related societal water problems, while at the same time ensuring an economically productive mining operation, is a precondition to ensuring that future mining operations will have a social licence to operate.

Those words are pretty strong stuff. If the minerals industry cannot find ways of resolving these problems, and resolving them in ways acceptable to the community, it is simply not going to be allowed to continue to operate. Such a threat would not have appeared even a dozen years ago.

Scope of the Book

One book could not cover all chains of production incorporated in the term "minerals industry." To create a coherent book, we decided to focus on *mining* as that term is commonly understood, which implies the use of a shovel—a hand shovel at one end of the scale and giants many stories tall at the other—rather than wells withdrawing liquids and gases. This kind of operation often involves relatively shallow but often vast open pit mining; or deeper underground activities centred on shafts that pierce aquifer formations, generate waste heaps the size of mountains, and discharge eerily beautiful toxic effluents. We also put to one side, for now, the huge volumes and equally large but different environmental and socioeconomic effects of the sand and gravel pits that appear ubiquitously near urban areas, as well as placer mining of alluvial ore bodies or extraction of riverbed sand and gravel.

Even within that editorial limitation, the impact of mining is too big to ignore in a world of oversubscribed water. The water that appears in statistics about mining use is only that small share directly used for operations, such as to cool drills, suppress dust, and to move material around inside the mine. Sizeable volumes of water are also required to maintain the community and the labour force it houses. Much greater volumes are affected by the very existence of the mine, whether it is on the surface diverting rivers and draining ponds or underground cutting across aquifers that need to be dewatered in order to make deep level mining safe. These often-ignored volumes of water do not appear in statistics about water use in mining, but they can be enormous, and they can make huge differences for people living and working in areas where mines coexist with earlier and typically more traditional ways of life.

From More Technology to Better Governance

For all the reasons highlighted above, it is inappropriate to see the water–mining nexus as a set of technical problems that need to be solved to facilitate more efficient mining. We cannot ignore mining's numerous social and environmental effects that are certainly local and often regional in scale. These characteristics imply that whatever the resolution, or absence of resolution, there will be winners and losers, and, further, that the problems are typically more political than they are technical. In this sense, they are very much like large dams. All of the chapters in this book recognize, and most focus on, the political components in the issues they explore.

When not properly regulated, mining can poison politics as much as, and in association with, poisoning land and water. In most countries of the world, corporate mining bodies must go through some administrative process to get a legal status that gives them permission to operate in a certain area, and to use certain resources, notably water, for some period of time and within some restrictions. This process almost inherently creates an incentive for the executives of the mining firm to establish a friendly relationship with the relevant power structure—certainly the local power structure and not uncommonly the national power structure—including both regulators and politicians. The process is smoothed by mining commonly taking place in remote areas whose inhabitants are already marginalized legally and politically.

Sadly, the mining industry has in many cases insisted on rights that were negotiated with governments that had little or no knowledge of, or sympathy for, local people or local conditions. In the worst cases, the companies have developed their own security forces or supported national security forces with loss of life on the part of community people or people who came to support their claims to land and water. Even where relations are "mutually beneficial" economically, local people are led to abandon traditional lifestyles and workstyles that were lower paying but sustainable in exchange for jobs in mines that are higher paying but more dangerous and less likely to be there over the long term.

Nearly half of the chapters in this book involve gold mining. Gold mining not only exemplifies but commonly highlights the problems. Contributing to the wide impact of the mining–water nexus is the tendency for gold to be exploited in a rush. Though extraction of gold ore is not notably different from other forms of mining, gold deposits tend to occur as veins or pods that, given the high returns per tonne mined, can be profitably exploited by relatively short-term ventures. Worse yet, the most common form of treatment is to extract the gold from the ore by use of cyanide. Both extraction by large earth-moving equipment and leakage from cyanide-laden ponds leave lasting scars on former agricultural or grazing lands, but little or no benefit for local communities. We also note that gold is often found in ore bodies that contain a range of other heavy metals, often discarded as waste where they are left to leach out into surface water and groundwater and blow away as dust.

No wonder either that all of the chapters in the book involve either the deficiencies of existing governance structures and the need for better ones, or explore the use of a new technique to identify and evaluate social and environmental impacts. In some cases, the needed changes are marginal; more commonly they are substantial.

Organization of the Book

The first four chapters in this book emphasize problems that could arise in almost any mining district in the world: unintended consequences of policy "reforms," efforts to increase data availability, and liabilities from both transboundary pollution and intraregional conflicts between water users.

Turton emphasizes that revenues from gold mining sustained the apartheid regime after 1961 in the face of economic sanctions and supported military spending. Among other efforts to preserve the mining industry, some water-based liabilities were nationalized. The unintended consequence was disinvestment from the industry, which triggered more nationalization and further disinvestment. Only with senior-level, science-based policy reform was South Africa able to cut the "Gordian Knot" created by the apartheid state in its self-imposed battle for survival. Water policy thus plays a critical but often invisible role in impeding the level of investment needed to rehabilitate mine-impacted aquatic ecosystems and landscapes.

Timms et al. take on the specific problem of hydrological measurement with specific reference to groundwater withdrawals in the Hunter Valley coal fields in Australia. Improvements in water reporting have enabled mine water issues at a watershed scale to be evaluated in unprecedented detail leading to improved security of water supplies for all water users. The authors highlight these major solutions in a region with increasing competition for water. They develop a new water use productivity curve and show that water take and use rates are unrelated to the type of mine or its rate of production. This suggests that site-specific practices and constraints are critical to improving how efficiently water is used.

In 2014, the United Nations Watercourses Convention (UNWC) came into force. As a signatory to the UNWC, South Africa is subject to its provisions including those that prescribe liabilities for transboundary pollution. Kinna asks whether acid mine drainage originating within South Africa and flowing through the Olifants/Limpopo River system into Mozambique might establish grounds for claiming liability. Whether or not such claims are actually pursued is another matter. However, Kinna concludes that the issue of liability and compensation for pollution of transboundary water resources is not going away anytime soon.

Turning from estimating water quantity, du Plessis and her colleagues focus on quantitative methods for projecting water quality changes as a result of dam construction. The catchment for the Grootdraai Dam on the Vaal River in South Africa is located within the economic and population core of South Africa. The authors found ways to quantify the intricate relationships between land cover and specific water quality parameters. The authors urge that future urban and mining developments not be approved in the absence of studies that project resulting changes in water quality in the region.

The next four chapters present case studies of mining–water problems in areas as diverse as South Africa, Peru, and Mexico. They explore a range of issues including the following: introducing unconventional resources in established mining districts, linkages between hydrological effects underground and social impacts on the surface, conflicts and power structures, and opportunities to bring local people into decision-making.

Esterhuyse and two colleagues present a pair of linked chapters about recent experience with unconventional oil and gas (UOG) projects around the world but with a focus on South Africa, which still has to embark on significant UOG resource development. In an initial study, they conclude that proper regulation of shale gas mining in South Africa is viewed as extremely important by decision-makers, and they go on to identify possible regulatory and monitoring tools to assist in governing this activity. The follow-up study three years later finds that little effort has been made to implement appropriate governance, which can be linked to fragmentation that is observed in governance structures. Therefore, the authors emphasize the need for increased attention to energy–water linkages, water–agriculture linkages, and water–human population linkages during the development of regulations for UOG development. They further emphasize that regulators must also address issues of scale linked to the institutional management of all the resources that may be affected during UOG extraction. Similar to the situation previously described by Kinna, the government has to be uncharacteristically forward-looking to avoid some very nasty, and foreseeable, situations.

Vela et al. explore linkages between hydrological effects from mining and water distribution to communities based on a case study of the Yanacocha mining district in the Cajamarca Region of Peru. They identify important concerns over changes in water flows, reduction of water table levels, and decrease of base-flows with impacts on people living in the same water basin. The authors question whether it makes economic or ecological sense to assign water rights without hydrological planning of the catchment and clear knowledge of the availability of water resources.

Stoltenborg and Boelens review a dispute over land and water rights near an open-pit gold and silver mine in Cerro de San Pedro, Mexico. They find that changes in land and water rights in Cerro de San Pedro result from a complex interplay among different actors, where the court systems, officials, and governments at diverse levels play a double and deeply troublesome role. To add to the problems, a multinational corporation used loopholes in the laws and its economic and discursive powers to serve its interests to the detriment of local communities. In addition, Stoltenborg and Boelens show how international agreements, such as the North American Free Trade Agreement (NAFTA), can have inequitable and unethical effects on the litigation process.

The remaining four chapters look for solutions to problems stemming from the mining–water nexus and expand the context from specific sites to mining as an international issue. Two of the four argue for wider use of selected tools, and a third looks at an international framework for understanding problems of governance. All three emphasize case study tests of applicability in regions as diverse as Ghana, Peru, and Suriname. The final chapter challenges our understanding of the word "reform," and whether mining reforms do more than just paper over structural problems.

Patrick and Bharadwaj start from the position that uncertainty over impacts to water quality from large-scale (formal) mining activity has raised human health concerns among campesino (peasant) communities in the Ancash region of the Peruvian Andes. They suggest that integrated water resources management (IWRM) might influence the outcome of industrial mining activities. Unfortunately IWRM is not practiced in Ancash to the detriment of the natural environment and highland communities. Mobilizing IWRM at the national level and scaling down IWRM implementation to the regional and local levels would, in their view, empower campesinos to join in water resources decision-making.

Danoucaras et al. ask how often mining-induced changes in water use and distribution turn out to be the source of social impacts. Their chapter explores the Social Water Assessment Protocol (SWAP), which is a series of questions on fourteen themes relating to how a community interacts with its water resources. A pilot study of the SWAP was conducted around a mining region in the Prestea-Huni Valley District in Ghana. The application showed that the SWAP has been well designed to capture the key elements of the social context of the region, and that it is an effective and systematic way to draw out the social issues of the communities that surround the mine.

Roy and her colleagues expand the exploration of solutions to mining–water problems in two ways. First, they point out that many areas affected by mining are also agricultural areas, so that the issue becomes one of mining and a water–energy–food nexus. Second, representing the secretariat for a newly formed Intergovernmental Forum on Mining, Minerals, Metals and Sustainable Development, they develop an analytical tool to bring out the interactions between mining and the water–energy–food nexus, and present a case study of that tool applied to the country of Suriname.

Finally, in a uniquely important perspective, Sosa and Zwarteveen look back at case studies of purported reforms in water governance in the mining industry of Peru and ask whether they are really improving governance or just neutralizing conflicts. They find that almost all conflicts are between multinational mining companies and local communities over the access to, control of and distribution of water. They then go on to show that, although legal and technical conflict resolution strategies are effective in temporarily defusing tensions, they do not address underlying political causes of conflicts. Instead, solving environmental conflicts involving large-scale mining operations requires explicitly admitting and dealing with the fact that these conflicts are always inherently political, situated, complex, and power laden.

Conclusion

Given the strong emphasis on the need for reforms in governance, the identification of pitfalls in even "enlightened" governance reforms in the past, and the number of chapters that found a balance of power that was heavily weighted against local people, attention to the reforms advocated in this book is more than academically advisable; it is politically essential.

For far too long, the dominant, closed-door paradigm in mining has been based on legal compliance only. This could be seen as the *legal licence to mine*, which typically involves little more than meeting minimum compliance levels and in corrupt systems, does not even go that far. This is no longer adequate. Growing recognition of the environmental and social externalities of mining, mediated by water, requires compliance with what could be called the *social licence to mine*. Governments and mining companies violate this licence at their peril.

Untying the Gordian Knot: unintended consequences of water policy for the gold mining industry in South Africa

Anthony Turton

Gold was central to the South African state from the outset. Its revenues sustained the pariah apartheid regime after 1961 in the face of economic sanctions and military spending. At that time, a regulatory regime arose that blurred the distance needed between regulator and regulatee. Water-related liabilities such as acid mine drainage were nationalized, burdening the post-1994 democratic government. Legal reform has sought to internalize those historic externalities through the application of the green-fields logic of global best practice. The unintended consequence is disinvestment, thereby hastening the nationalizing of all remaining liability. A new approach is needed.

Introduction

Policy and legislation tend to follow progress in industry and the economy. This is particularly true in an emerging extractive economy, where diversification has not yet emerged after the depletion of the ore bodies that originally sustained mining. What happens when the conditions on which policy and legislation are premised change? Can policy and legislation guide a transition from a mining-based economy to a post-mining beneficiation economy? In particular, what role does water policy – specifically the retrospective application of the polluter-pays principle – play in decisions on investment into the mining industry? What are the consequences for long-term rehabilitation of mine-impacted aquatic ecosystems?

This set of issues is currently at the heart of the gold mining industry in South Africa. Once the richest gold producer in the world, supplying a staggering 40% of all gold ever produced in recorded history (Hart, 2013), the Witwatersrand Goldfields are now in rapid decline. Johannesburg is one of the few cities in the world that is not on a river, a lake or a seafront. It straddles the continental divide between two major transboundary rivers in southern Africa – the Orange (discharging into the Atlantic Ocean) and the Limpopo (discharging into the Indian Ocean). Its very existence is rooted in gold mining, and it has grown to become the financial hub of continental Africa (Turton et al., 2006). All of this is changing rapidly, however. The majority of the resource is mined out, and three of the four major groundwater basins are flooded or in an advanced stage of flooding[1] (Coetzee, Hobbs, Burgess, Thomas, & Keet, 2010). One consequence is acid mine drainage, a flow of highly acidic and heavy metal–laden water that has been decanting unabated from the Western Basin since 2002. A major engineering race is currently underway to prevent a similar environmental catastrophe from happening in the Central and Eastern Basins of the Witwatersrand Gold Mining Complex.

Although the focus here is on gold, an even more formidable problem looms and must be noted in passing. The city of Johannesburg is famous for its mine dumps, often featuring on post-cards sent by tourists back home. Lurking within those dumps is a staggering 600,000 tonnes of uranium, discarded as waste over the last 135 years of mining (Winde, 2006). This uranium is now being mobilized into the headwaters of these two transboundary river basins by rainfall (Coetzee, 1995; Coetzee, Wade, & Winde, 2002a, 2002b, 2005; Camden-Smith, Pretorius, Camden-Smith, & Tutu, 2015) and distributed over a wide area by dust storms. Little is known regarding the consequences of this for all stakeholders at different scales (Turton, 2014), and water policy makers are confused.

So, what can be done to avert an economic and social catastrophe brought on in large part by environmental policy in general and water resource policy (both surface and underground) in particular? This article explores the complexity associated with policy formulation under conditions of contestation, data uncertainty and high levels of risk. The first part gives a historic overview of the evolution of water policy and mining in South Africa to provide context. The second part presents data on the current state of the gold mining industry in South Africa. The third part analyzes the current policy approach and concludes that the failure to distinguish between environmental management in green-fields and brownfields developments destroys the business case for the latter, hastening the demise of the industry, and preventing funding of rehabilitation initiatives in mine-impacted aquatic ecosystems.

Overview of the history of mining in South Africa

It's no secret that South Africa's once-massive gold mining industry has been shrinking for years, leaving billions of dollars in processing plants, worker housing, and pipelines at risk of demolition and the scrap yard. Seemingly keen to stave off this scenario, the government wants mining companies to find alternative uses for idled facilities to support communities and even continue to provide jobs. "When we talk about mine rehabilitation we are not talking about just putting back the sand, but actually rebuilding those communities," Mines Minister Ngoako Ramatlhodi told Reuters. [This] ... presents a rosy picture of Harmony Gold ... transforming one of its disused leaching facilities into a biofuel-producing plant, but we can't help but feel a little uneasy as to what the above statement will mean for the industry as a whole. Environmental rehabilitation is a good thing, but tough enough – community rehabilitation is a tall order. SA's mining problems are far-reaching and difficult to surmount. In our view, the government should take every care not to burden the already overburdened industry – risking its very survival. (International Spectator, 2014 – this is a confidential newsletter subscribed to by investors)

To understand this bleak assessment of the present, it is useful to review the evolution of water and environmental policy in South Africa as it pertains to the gold mining industry. The Second Anglo-Boer War was driven by Great Britain's desire to gain control over the largest known deposit of gold in the world (Pakenham, 1992). Due to the dogged persistence of the fighters in the two Boer Republics, who refused to capitulate even after their capital cities had been taken, a scorched-earth policy was introduced (Pretorius, 2001) that involved the use of concentration camps for the wives and children of the combatants (Fawcett, 1901; Krebs, 1992). The modern state of South Africa created by the Act of Union in 1910 was an amalgam of the two vanquished Boer Republics (Transvaal and Orange Free State) and two former British colonies (Cape and Natal), with a deeply entrenched collective memory of violence and inequality (Meredith, 2007).

In this context there have been five distinct phases of policy evolution (mineral, water and environmental), with three clear policy monopolies interspersed by periods of intense policy instability. This is presented graphically in Figure 1. On the horizontal axis of Figure 1 time is represented, with distinct periods of historic significance recorded as pivotal moments in the evolution of policy. There have been five pivotal events that have triggered five distinct policy paradigms (Turton, 2009). These are summarized briefly as follows.

Policy Paradigm I commenced with the formation of South Africa as a single sovereign state rising from the ashes of war (Meredith, 2007; Morris, 1971). It spanned the period from 1910 (Act of Union) to 1948, year of the first general election after World War II. Its key defining characteristic was British hegemony focused on the extraction of gold and repatriation of the wealth so created back to the United Kingdom. All of the original legislation that gave rise to the various government departments that played a role in the regulatory aspects of mining from that date forth served these purposes. There was no emphasis on water or environmental management at this time; Policy Monopoly I was based on simple extraction. This period was characterized by growing discontent by certain Afrikaner (Boer) political elites, all of whom had living memory of the British concentration camps, and collectively wanted to regain the freedom lost during the Anglo-Boer War (Liebenberg, 1987). Running parallel to this process was a growing political consciousness among the disenfranchised black majority (Karis & Carter, 1972; Liebenberg, 1994; Mbeki, 1984).

The significance of Policy Paradigm I from a water resources perspective is that the main focus was on the building of hydraulic infrastructure to act as a foundation for the growing economy, with no emphasis on water quality.

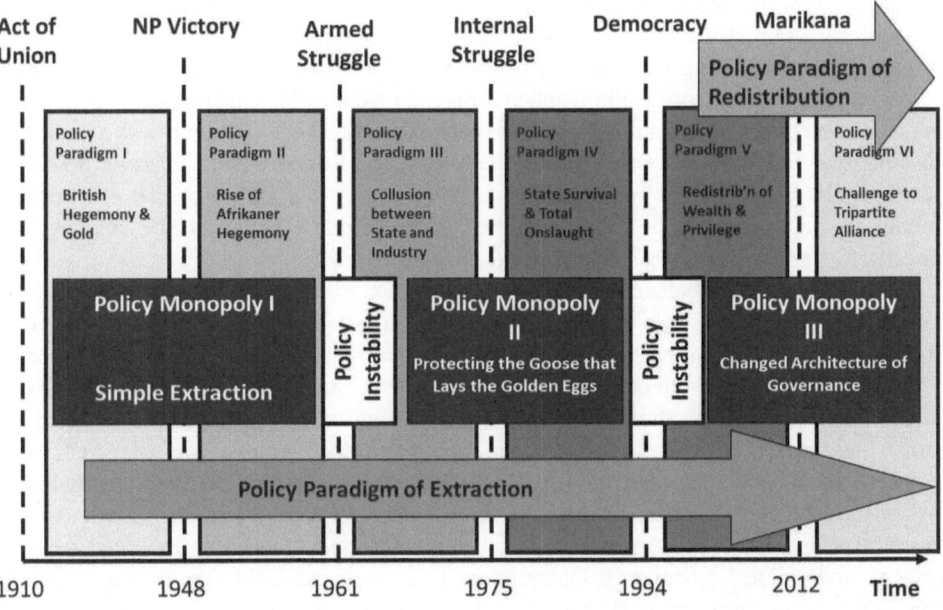

Figure 1. The evolution of policy as it relates to water and environmental management in the mining sector (after Turton, 2009).

Policy Paradigm II occurred from 1948 to 1961, after the Nationalist Party won the first election in the post–Boer War era (Meredith, 2007; Turton, 2009). This did not change the policy focus significantly, other than transferring political reins back into Afrikaner hands. Water policy was encapsulated in the Water Act (1956), which focused on building infrastructure for irrigated agriculture and meeting the needs of a newly industrializing economy (Turton, Meissner, Mampane, & Seremo, 2004). While this did require mining companies to take measures to prevent pollution arising from their activities, the impact of that pollution on water resources was not yet quantified.

These first two policy paradigms ignored environmental and social issues, and can be lumped together as Policy Monopoly I: simple extraction. Water policy was primarily driven by the need to build infrastructure; water quality was decidedly a secondary concern, in part because at that time it was still of a relatively high standard.

The Sharpeville Massacre of protesting black citizens in 1960 led to international isolation of South Africa for its policy of racial segregation (apartheid). This included expulsion from the British Commonwealth and eventually pariah-state status under comprehensive economic sanctions. This played a significant role in the evolution of subsequent policy.

Policy Paradigm III thus emerged from a period of intense policy instability under-pinned by a major loss of investor confidence and the withdrawal of foreign investment on a significant scale. Central to this was the restoration of investor confidence by means of the Jordaan Commission of Enquiry (Turton et al., 2004), which opened up the deep-level reef previously considered too dangerous to mine because of the perched dolomitic aquifer, located in the Far Western Basin (Jordaan et al., 1960). In effect between 1961 and 1975 the state ceased to be an independent regulator of the mining industry, as both parties increasingly colluded to boost profits – and thus taxes to the state – by externaliz-ing liabilities as far as possible. This gave rise to Policy Monopoly II, which was about protecting the goose that laid the golden eggs.

The significance of Policy Paradigm III from a water resources perspective is that this period of time witnessed the aggressive phase of the national hydraulic mission (Blanchon, 2001; Turton, 2000; Turton et al., 2004). Major investment poured into water infrastructure, with a view to developing the economy in the face of international isolation by developing the coal and energy sectors while using the revenue from gold. Oil-from-coal technology was pioneered as a strategic initiative under the brand name SASOL; and it was also during this time that ESKOM (the national electricity producer) underwent a massive period of growth, underpinned by the inter-basin transfer of water as a matter of national priority, without regard to impacts on other countries in the region (Basson, 1995; Basson, Van Niekerk, & Van Rooyen, 1997; Blanchon & Turton, 2005; Heyns, 2002).

In 1975 the regional balance of power changed when Portugal announced its withdrawal from the wars of liberation in Portuguese-speaking Africa. This made the Rhodesian Bush War unwinnable, with a second front opening up in Mozambique (Martin & Johnson, 1981), triggering a ripple effect in Angola (Venter, 1975). South Africa became embroiled in various local wars of liberation (Turner, 1998), draining massive sums of money from the fiscus. Significantly, South African military involvement in Angola was triggered when engineers working on the Calueque hydroelectric project (part of the national hydraulic mission) were taken hostage. This triggered a policy response from the South African government based on the concept of a 'total onslaught' and focused on the need for state survival, underpinned by military intervention, in a perceived Cold War theatre of opera-tions against the post-colonial Soviet-backed regimes in Angola and Mozambique (Gutteridge, 1983). Manifested as the need for a 'total strategy' based the writing of a French strategist at that time (Beaufre, 1965; Brodie, 1965), this became embedded in the water sector as part of the Lesotho Highlands Water Project (Blanchon & Turton, 2005; James, 1980). Policy Paradigm IV was thus about state survival in the face of a perceived total onslaught in a localized theatre of the Cold War (Geldenhuys, 1984).

The significance of Policy Paradigm IV from a water resources perspective was that yet again, there was no significant place in this policy framework for the environmental and social concerns that were starting to manifest in the mining sector. Water was increasingly seen as a strategic issue, with major interventions in the form of supply-side engineering, often on transboundary rivers, in which inter-basin transfers played a key role (Basson, 1995; Basson et al., 1997; Turton, 2000; Turton et al., 2004). There was one notable new development in water policy at this time. As a direct response to the deteriorating national security situation in 1975, attention was again drawn to the mining industry as a key element of state survival during the Cold War. A significant policy reform took place in late 1975 when the Fanie Botha Accord[2] – named after the minister of water affairs at that time – was negotiated between the gold mining industry and the government regulatory authority. The following became key elements of water policy for the mining sector:

- Pollution control measures abandoned by mining companies prior to 1975 would become the responsibility of the state.
- Where mines owned the land on which such pollution had occurred, they would facilitate the transfer to the state, effectively nationalizing the liability accrued prior to 1975.
- Where the mining company owned the mineral rights but not the land, it would be required to assist the state in acquiring that land for purposes of consolidating the liability, effectively taking it off the balance sheet and nationalizing it.
- Where mining had ceased before the promulgation of the 1956 Water Act, and had disposed of its mineral rights, then the company was *not* required to assist the state in acquiring land for pollution-control structures.
- Provided that mining companies had taken steps to control pollution based on the 1956 Water Act, the state would assume management of all pollution-control works after mining had ceased.
- When mining ceased on specific land, the mining company had the right to develop that land for other purposes, with the management of pollution control arising from mining reverting to the state, provided that such measures had been approved by the state during operations. Pollution-control measures arising from the non-mining activities would then become the sole responsibility of the mining company, the state assuming full liability for the mining-related pollution.
- Should a mining company wish to return to old mining operations for whatever reason after the state had assumed liability, then the company would be held liable only for pollution-control measures arising from the new activities.

In effect, then, the Fanie Botha Accord nationalized all liabilities prior to 1975, on the condition that pollution-control devices had been installed and approved per the 1956 Water Act. This enabled revenue to the embattled state to be maximized by taking significant liabilities off the collective balance sheet of the mining companies (see Figure 2). Policy Monopoly II thus consolidated Policy Paradigms III and IV, centred solely on the protection of the gold mining industry, by externalizing and nationalizing all environmental liabilities as a perceived means of state survival. This gave rise to the period known as the Midas Touch, in which South Africa was able to counter growing internal insurgency while fighting a series of external wars (Geldenhuys, 1983, 1984)

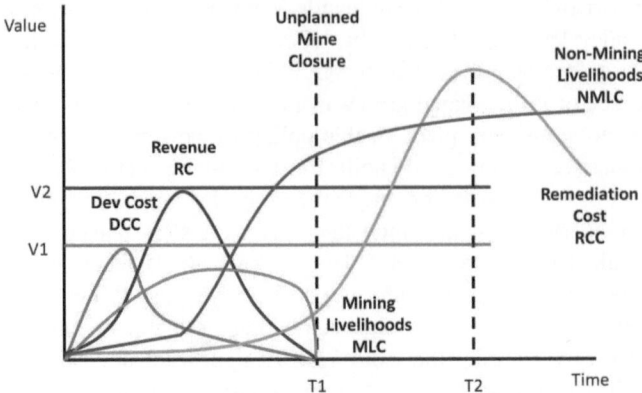

Figure 2. Economic model showing the flow of money over time from a given mining site, by specific source of cost or revenue. Policy Monopoly II (see Figure 1) saw the removal of all significant environmental liabilities from the balance sheet in order to ensure state survival by maximizing taxable revenues. This means there is insufficient money set aside to fund post-closure rehabilitation of mine-impacted aquatic ecosystems.

using the minerals complex as a core revenue source (Gutteridge, 1984). In effect the state ceased to regulate the gold mining industry, instead allowing self-regulation without democratic oversight, in exchange for a greater flow of funding to meet the state's increasingly dire financial requirements.

Policy Paradigm V was triggered by an intense period of policy instability associated with the final days of the armed struggle in which South Africa nudged close to a fully fledged civil war. The pivotal moment came when South Africa adopted a democratic constitution in 1994. This triggered a new paradigm of redistribution of wealth and privilege under Policy Monopoly III, which was part of a changed architecture of governance. Tension now exists between the old policy paradigm of extraction and the new policy paradigm of redistribution. More importantly, emphasis is now placed on state capacity as a regulator, without the institutional memory needed to underpin that process. The Marikana crisis[3] in August 2012 resembled the policy instability associated with the Sharpeville Massacre in 1961. This has potentially given rise to an as yet ill-defined Policy Paradigm VI called Challenge to the Tripartite Alliance. The Tripartite Alliance consists of the African National Congress (ANC), the South African Communist Party (SACP) and the Confederation of South African Trades Union (COSATU). The emergence of a highly militant but non-aligned trade union (Association of Mineworkers and Construction Union – AMCU) during the Marikana incident has triggered the disintegration of the Tripartite Alliance that played a pivotal role during the democratic transition (COSATU, 2014). In the absence of robust state capacity to effectively apply laws and policies, as well as to provide strategic leadership, in an environment where critical decision making occurs at the company level, underpinned by the inability to make adaptive responses in the form of policy reform, this development is now accelerating the demise of the industry.

The significance of Policy Paradigm V from a water resources perspective was the nationalization of all water-related liabilities in return for increased revenue flows to the fiscus. The unintended consequence of this relates to the application of the polluter-pays principle, along with other environmental management principles, in a retrospective manner that now manifests as a constraint on investment in highly impacted aquatic ecosystems draining brownfields sites in critical need of rehabilitation.

Understanding Policy Monopoly II – the externalization-of-costs model

An analysis of these policy dynamics was done by the principal author and his team at the Council for Scientific and Industrial Research in 2006–2007 (Adler et al., 2007a) as part of the Mining, Minerals and Sustainable Development programme. From this a model was developed (Figure 2) to illustrate the economics of gold mining in South Africa (Adler et al., 2007b).

On the vertical axis we have value, with time represented on the horizontal axis. The Development Cost Curve (DCC) represents the cost of developing a given mine, peaking at value V1. The Revenue Curve (RC) is out of phase with the DCC, representing the flow of cash after the mining infrastructure (shafts, processing plant etc.) has been developed, peaking at value V2. The profit of the given mining operation is the area beneath the RC minus the area beneath the DCC. The revenues arising from mining in the form of job creation are shown as the Mining Livelihoods Curve (MLC). This includes the wages accrued from all mining-related activities, both on and off the site. Eventually the RC and DCC reach a nil value and mining ceases to be economically viable (T1). This is called mine closure and is characterized by a dramatic fall in wage remittance (MLC) arising from the cumulative effect of the collapse of mining-related employment. This causes catastrophic social collapse in the area around the mine, often associated with the decline in the economic viability of many small towns, accompanied by an increase in crime as alternative livelihoods are sought. These brownfield sites almost always manifest with severely degraded aquatic ecosystems in critical need of rehabilitation.

When mine closure occurs (T1) it does not necessarily mean that the resource is totally depleted, merely that the cost of extraction exceeds the potential value *under the prevailing set of regulatory conditions*.[4] At this time the environmental liabilities, shown here as the Remediation Cost Curve (RCC), are not yet fully manifest. These increase rapidly after mining stops and the pollution-control devices revert to the state (per the 1975 Fanie Botha Accord), peaking at an unknown value (>V2) at an unknown time in the future (T2). In essence, then, the externalization-of-costs model was deliberately introduced by an embattled pariah state between 1975 and 1994 (Figure 1), enabling mining companies to show substantial profits by removing all environmental (and thus social and economic) liabilities from their balance sheets, but now manifesting as a constraint on future economic development in the post-mining era. The Non-Mining Livelihood Curve (NMLC) was not considered by the regulator at the time; it will be dealt with later in this article.

The water and environmental policy challenge inherent to Policy Paradigm VI is thus centred on the need to transition from an extractive to a post-mining beneficiation type of national economy in which externalized environmental liabilities now manifest as constraints on investment and job creation. More importantly, insufficient capital has accrued to fund rehabilitation, so attempts by the state to re-internalize those liabilities to make up this financial shortfall act as a disincentive to new investment by destroying the business case for brownfields operations. The retrospective application of the polluter-pays principle is thus inappropriate as a policy instrument if the rehabilitation of mine-impacted aquatic ecosystems is to be viable.

Demise of the gold mining industry in South Africa

Noting the evolution of policy over time, it now becomes instructive to analyze the performance of the gold mining industry with a view to assessing its remaining useful life. More importantly, it is necessary to determine when gold mining will cease, in order to better understand the water and other policy reforms needed to ensure a soft landing in the transition from an extractive economy to a post-mining economy in which historic environmental externalities now manifest as constraints to investment, aquatic ecosystem rehabilitation and job creation.

A statistical analysis of Witwatersrand Goldfields production (Hartnady, 2009), based on Hubbert Theory[5] as used by Campbell and Laherrère (1998), was used to inform the Gauteng Department of Agriculture and Rural Development assessment of the policy implications arising from the many mine residue areas (MRAs) around Johannesburg (GDARD, 2011). These MRAs are home to around 1.6 million people, mostly living in informal settlements on the 5445 ha of hazardous mine residues that could be rehabilitated (Tang & Watkins, 2011). This analysis is presented as Figure 3.

From this assessment it is evident that production peaked in 1970, with a rapid near-linear decline subsequent to that. There are three discrete sub-cycles. The first, peaking in about 1930, was driven by shallow mining in the Eastern, Central and Western Basins

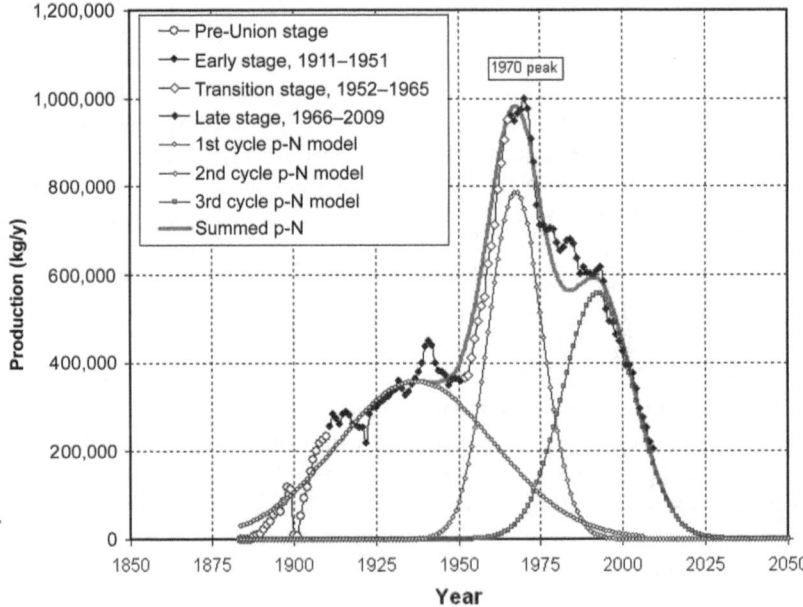

Figure 3. A statistical analysis of production from the Witwatersrand Goldfields reveals three discrete sub-cycles (GDARD, 2011; Hartnady, 2009). The third sub-cycle could create a soft landing by extending the life of marginal mines, provided that policy reforms are implemented.

Note: The "p-N Cycle" is a statistical analytical tool developed by Hubert in 1956 that shows the rate of depletion of a natural resource over time. N refers to the resource in question and p refers to the peak production of that resource with a given technology. In the case of gold production in South Africa there are three distinct p-N curves, each associated with different phases of technological development. The same holds true for oil where hydraulic fracturing has created a new p-N curve for shale previously considered to be uneconomic to recover.

(Davenport, 2013). The second peak was driven by deep-level mining, most notably in the Far Western Basin, after the decision was made by government to dewater the massive dolomitic aquifer that was perched above the reef (Jordaan et al., 1960) to accelerate economic growth during Policy Paradigm III. The third peak is not well defined. It is driven by improvements in metallurgy that enable gold to be extracted from old tailings dams, along with improvements to engineering processes that allow safe pillar extraction (Grice, 1998), manifesting as a brief interruption to the otherwise linear decline. This could become significant if the necessary policy reform takes place as suggested later in this article. A central characteristic of this analysis is the sharp decline in production during Policy Monopoly III.

In this regard the senior gold analyst working for the Old Mutual Gold Fund Manager stated the following on 17 June 2014:

> Predicting the future of gold mining in South Africa has become easy.... Having peaked in 1970, it has fallen to a 109-year low of 167 tons in 2012.... For exactly 20 years now [i.e. coinciding 100% with Policy Monopoly III], SA's declining output has followed a distinct linear pattern (if we disregard the effect of the 2012 strike). In itself, this is highly remarkable since it has completely disregarded the law of price elasticity: no supply response at all following bullion's spectacular price increase after 2001.... After more than 135 years ... SA is likely to hoist its last skip of gold-bearing ore from the once giant Witwatersrand deposit in 2019.... Some 130,000 direct jobs will be lost, with many more disappearing on the periphery in goods and services. By 2020, some R25bn of earnings annually will have stopped flowing to employees. Sadly, the poor will be particularly hard hit in the labour-supplying areas, where mineworkers tend to have many dependants. (Schroder, 2014)

This problem has been exacerbated by five significant drivers that act as significant constraints on water policy implementation.

First, mine workers are extremely well organized, this having been a central feature of the struggle for democracy in South Africa. They have successfully managed to negotiate wages significantly higher than peer cohorts in other jurisdictions. The living wage demanded for rock drill operators by the militant AMCU[6] (Association of Mineworkers and Construction Union) during the 2012 Marikana strike of ZAR 12,500 per month (ZAR 10 = USD 1, approximately), which broke the Tripartite Alliance (Policy Paradigm VI), is far above similar cohorts in Russia (ZAR 1,666), China (ZAR 2,184), Zimbabwe (ZAR 2,406) and Peru (ZAR 2,839) (Schroder, 2014). Since in 2012 South African mines produced 39% less gold per worker than they did in 2002 (Schroder, 2014), wages have been systematically decoupled from productivity (see the MPC 2 trajectory in Figure 5) and are now acting as a significant disincentive to foreign direct investment. This is directly reflected in the Fraser Institute survey of mining jurisdictions (Table 1), where South Africa now ranks 109th on labour relations out of the 112 jurisdictions listed, with an astonishing 17% of respondents saying they would not invest in South African mining due to the current state of labour relations (Wilson & Cervantes, 2013).

Second, policy reform in the environmental field has been radical in the post-1994 transition to democracy. In effect, the historic legacy of mining, centred on the deliberate regulatory approval of the externalization of environmental and other liabilities during Policy Monopoly II (1961–1994; see Figure 2), has been reversed. All mining operations are now expected to internalize historic externalities by paying a significant cash deposit up front (NEMA, 2014). The polluter-pays principle, enshrined in the National Water Act and accepted as international best practice, is increasingly the rallying cry of anti-mining activists, relentlessly applying pressure on old and marginal mining operations that are

Table 1. Summary of a survey of mining jurisdictions as rated by industry executives.

Top four categories[a]			
	% score	of 112	Quartile
Policy perception index	39.8	64	1 to 28
Best practice mineral perception	65	37	29 to 56
Investment attractiveness	54.7	53	57 to 84
Current practice mineral potential	37	55	85 to 112
Room for improvement	30	n/a	n/a

Fifteen questions[b]					
	A	B	C	D	Best jurisdiction
Uncertainty concerning the administration, interpretation and enforcement of existing regulations	72	85	4	58	Western Australia
Uncertainty concerning environmental regulations	53	69	8	30	Namibia; Wyoming, USA
Regulatory duplication and inconsistencies (including federal/provincial and interdepartmental overlap)	69	83	2	33	Eritrea
Legal system (fair, transparent, non-corrupt, timely, efficiently administered, etc.)	64	63	6	50	New Brunswick, Canada
Taxation regime (includes personal, corporate, payroll, capital and other taxes and complexity of tax compliance)	53	72	6	38	Alberta, Canada
Uncertainty concerning disputed land claims	70	97	2	38	Ireland
Uncertainty about which areas will be protected as wilderness, parks or archaeological sites	45	57	10	25	Burkina Faso
Quality of infrastructure (including access to roads, power availability, etc.)	26	32	22	60	Finland
Socio-economic agreements/community development conditions (includes local purchasing, processing requirements, supplying social infrastructure)	53	73	6	47	France
Trade barriers – tariff and non-tariff barriers (restrictions on profit repatriation, currency restrictions, etc.)	43	63	17	51	Sweden
Political stability	72	85	8	75	Alberta, Canada
Labour regulations, employment agreements and labour militancy/work disputes	81	109	0	45	Sweden
Quality of geological database (includes quality and scale of maps, ease of access to information, etc.)	23	42	23	78	Ireland
Security situation (including physical security due to threat of attacks by terrorists, criminals, guerrillas, etc.)	59	83	4	85	Minnesota, USA
Availability of labour and skills	45	66	24	63	Ireland

Notes: "% Score" refers to the number of respondents in that specific category expressed as a percentage of total respondents. "of 112" refers to the number of respondents in each specific category out of a total of 112 polled. In the percentage 112 = 100%.
a. These are general perceptions: an overall view on policy; attractiveness of minerals; applying best policies; and overall investment attractiveness (weighted 40% policy perception and 60% mineral resources).
b. The 15 questions were asked of mining executives around the world. Responses are on a scale of 1–5, from factor encourages investment to would not invest. Column A is the percentage of respondents answering 3 (mildly deters investment), 4 (strongly deters investment) or 5 (would not invest due to this factor). Column B is the ranking out of 122 jurisdictions for this factor. Column C is the percentage of respondents answering 1 (this factor encourages investment). Column D is the percentage of respondents in the top-rated jurisdiction for this factor answering 1 (this factor encourages investment).
Source: Wilson and Cervantes (2013).

simply incapable of reversing the legacy of Policy Monopoly II. This failure to distinguish between greenfields (with 100% of the mineral resource available to fund rehabilitation) and brownfields operations (with insufficient economically viable mineral resource left under prevailing regulatory conditions) is a critical policy flaw manifesting as an investment constraint.

Third, the acid mine drainage issue that burst into the public arena when decanting to the surface was reported by the media in the Western Basin in 2002 has given rise to a growing narrative of blame-seeking (Coetzer, 2008; Noseweek, 2009; Segar, 2013) that has an inherently high conflict potential (Botha, 2013). Increasingly angry public opinion, moulded by anti-mining activists, makes no attempt to understand that the current situation is a manifestation of a policy failure of Policy Monopoly II and not the deliberate intention of a seemingly reckless mining industry. Of even greater concern is the growing evidence that points towards uranium contamination as a largely unquantified but significant risk that is just beginning to be recognized (Camden-Smith et al., 2015; Coetzee, 1995; Coetzee, Winde, & Wade, 2006; Toffa, 2012; Turton, 2014; Wade, Woodbourne, Morris, Vos, & Jarvis, 2002; Winde, 2010; Winde & van der Walt, 2004).

Fourth, the only significant land left to develop in and around the city of Johannesburg is located on MRAs (GDARD, 2011). The uncontrolled settlement of around 1.6 million people is a growing risk, given the toxic nature of the mine tailings and the geotechnical instability caused by shallow undermining from illegal artisanal miners (Tang & Watkins, 2011; Toffa, 2012). This means that the government has a tough decision to make: move the hazard away from the people; or move the people away from the hazard (GDARD, 2011). The latter is unlikely, given South Africa's experience with forced removals, leaving the rehabilitation of MRA land the only viable option, unless the state is willing to assume this responsibility as mining companies inevitably become insolvent. Given the experience in the case of the Tudor Shaft (SERI, 2013), a small uranium-rich dump on the West Rand that has seen environmental activists oppose all attempts at implementing both options (Bega, 2012; Segar, 2013), the prognosis for rehabilitation of mine-impacted aquatic ecosystems seems bleak as the overall policy framework currently stands.

Finally, the collapse of the formal mining sector has given rise to a burgeoning illegal artisanal mining community, currently accounting for the loss of a staggering ZAR 5 billion per annum (Hart, 2013; Wolmarans, 2014). This is also closely associated with sophisticated criminal syndicates that use the gold bullion for money laundering on an international scale. As mines become insolvent, they are stripped out by gangs that attack the remaining mineral resource, which is large enough to sustain informal mining for more than a century. The unintended consequences of this have been revealed by Grootvlei[7] and Blyvooruitzig,[8] both of which resulted in a sharp increase in criminal activity once formal mining ceased to be economically viable and the failure of the mining company caused unplanned closure (T1 in Figure 2).

Unintended consequences of the current policy approach

The outcome of these five significant drivers presents as a classic dilemma confronting the government as well as invested mining companies. The policy response thus far is not informed in any way by the historic legacy of mining or the unintended consequences arising from Policy Monopoly II (Naidoo, 2014). In the absence of evidence-based policy reform, attempts by the state to accrue the capital needed for post-closure rehabilitation have incentivized the acceleration of the nationalization of remaining liability by discouraging investment into brownfields operations. The aggressive application of the polluter-pays principle, without taking cognizance of this complex legacy, amounts to the retrospective implementation of policy that is simply unworkable if the rehabilitation of mine-impacted aquatic ecosystems is to attract the magnitude of investment needed.

An economic model capable of asking the what-if questions needed for policy analysis is shown in Figure 4. Two monetary values are represented vertically. On the left-hand axis the comparative value at any given time is shown as either a positive or a negative quantum, with 10 arbitrarily selected as the peak, equivalent to V2 in Figure 2. On the right-hand axis the cumulative value is expressed as a percentage of the total cost of remediation, representing the RCC in Figure 2. CV1 represents the legally required cash deposit to cover the cumulative environmental liability that was nationalized during Policy Monopoly II should a company wish to invest in the brownfields site today. At zero cumulative value there is 100% of the mineral resource in the ground and no pre-mining work has been undertaken, with no liability at the start of mining. At 100% cumulative value there is zero mineral resource that can be viably extracted from the ground from which to fund rehabilitation, thereby representing the full cost of remediation required at that point. The important point is that when T1 occurs, for illustrative purposes there is still some of the mineral resource left in the form of crown, shaft and stope pillars, as well as residual gold left in tailings from older, less efficient metallurgical processes. Unplanned closure occurs *only* when the commercial viability ceases, based on current regulatory assumptions and the social cost structure. The Cumulative Rehabilitation Requirement Curve (CRRC) is the quantum of funds required to rehabilitate the impacted land and water resources at that particular time in the mining project. Thinking about it as

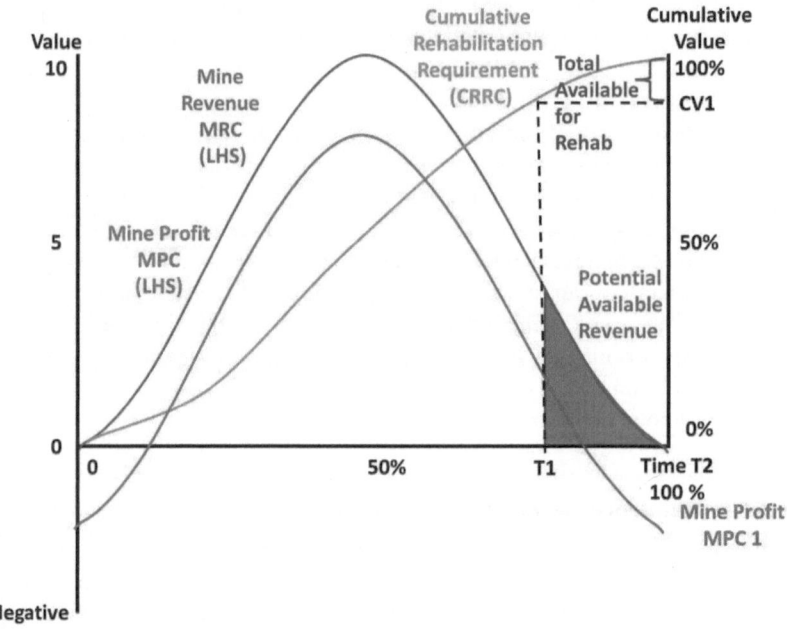

Figure 4.   Model showing the unintended consequences of Policy Monopoly II.
Note: LHS, left hand scale.

vertical lines, the rehabilitation is 100 (as it is accumulative) where the revenue bell curve is cut into each unit of time; and the vertical lines bisecting the bell curve summed to give this value. It is thus a multiple of the total rehabilitation amount, and the revenue needs to cover capital expenditure (CAPEX), operational expenditure (OPEX), rehabilitation costs and profit for there to be a viable project.

Once again the significance is that at T1, the majority of the resource has been extracted, but there is still some remaining on which future rehabilitation can be based, provided that the policy response accepts that the remaining resource is insufficient to carry the accrued liability of all past mining.

On the horizontal axis time is expressed as a percentage needed to totally deplete the mineral resource as a viable mining operation from an original pristine pre-mining value. T1 represents the moment when the regulatory regime changes the economic viability by materially altering the cost structure. In the case of the Witwatersrand Goldfields, this happened at the start of Policy Monopoly III in an attempt to overcome the shortcomings of Policy Monopoly II. Importantly, this does not represent the total depletion of the mineral resource. The Mine Revenue Curve (MRC) in Figure 4 is identical to the RC for a specific mine (Figure 2) or the bell curve shown for the entire industry in Figure 3. The Mine Profit Curve (MPC 1) represents the difference between the area under the RC and DCC shown in Figure 2.

Seen in the context of the historic evolution of mining policy shown in Figure 1, most notably in light of the Fanie Botha Accord that was a feature of Policy Paradigm IV and Policy Monopoly II, insufficient capital has been set aside to fund post-mining rehabilitation of the magnitude defined by the value of the CRRC. If one applies that retrospectively, at the time of mine closure (T1), when the majority of the mineral resource has been depleted and what remains is marginal at best – as required by the NEMA (2014) amendment – to fund 100% of the accumulated liability (CRRC), then it simply destroys the business case by making such a venture unattractive to any investor. This drives the inevitable outcome of insolvency for all remaining marginal mines, consistent with the non-elasticity of the supply curve in Figure 3 highlighted by Schroder (2014). The unintended consequence of this is that the rehabilitation of mine-impacted aquatic ecosystems and landscapes naturally reverts to the state, with its limited technical capacity and insufficient funds with which to accomplish this complex task.

This need not be the case, however, given that at T1 there is still some of the mineral resource left in the ground. If policy reform is initiated, then the case shown in Figure 5 is possible.

If one accepts that when unplanned closure occurs (T1) there is still a quantum of the mineral resource remaining, then the potential available revenue to fund the shortfall arising from the Fanie Botha Accord is represented by the area below the MRC and above CV1 between T1 and T2. The revenue available to rehabilitate brownfields sites is reflected in the triangle shown as Potential Revenue beneath the MRC. There is no relationship between the area under this portion of the curve and the revenues needed for full rehabilitation. The revenues needed are the value of the CRRC at any given point in time and are naturally small under such circumstances, as suggested by the tiny triangle beneath MPC 2 in Figure 5. While the total available for rehabilitation is small, it not insignificant when one considers the harsh consequences of the alternative – a nationalized liability with limited state capacity to rehabilitate (Turton, 2014). Profits are going to be materially smaller as the differential between the MRC and MPC is OPEX and any further required CAPEX. (Note: the curves are not to scale for illustrative purposes.) This is where the minister's comment noted above becomes relevant. If mining is to effectively become part of the solution by

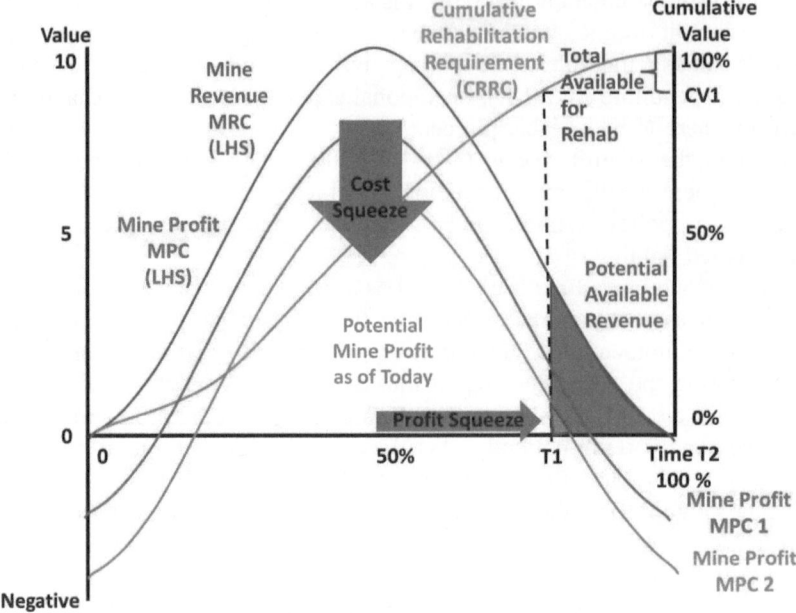

Figure 5.   Model showing that sufficient profit can attract direct investment into brownfields sites, provided that policy reform differentiates this reality from greenfields logic.

rebuilding communities (International Spectator, 2014), then the only possible way is to unlock the potential found in non-mining livelihoods (NMLC in Figure 2). The way this can be accomplished is to incentivize the following through policy reform:

- Distinguish greenfields logic (assuming a new development with a large ore body and a relatively unimpacted environment) from brownfields circumstances (with a depleted ore body and a highly impacted environment) where 100% of the liability has been accrued but only a limited amount of the mineral resource is left to fund possible environmental and social rehabilitation.
- Create a relatively uncontested vision for a post-closure landscape and environment for use as a blueprint for potential rehabilitation (Juwet & Lyssens, 2014).
- Quantify the cumulative impacts of mining in a given hydrological management unit that need to be mitigated by an integrated closure strategy (Mudd, 2007; Van Tonder & Coetzee, 2008).
- Encourage investment into brownfields sites by allowing for the quantification of benefits arising from the removal of hazards such as surface tailings dams, unmitigated acid mine drainage flows, geotechnical instability caused by shallow undermining, and accumulation of uranium and other metals in wetlands, as defined by the vision.
- Once these potential benefits have been quantified, bring them onto the balance sheet in a transparent way to offset the liabilities currently acting as a disincentive to investment into brownfields sites.
- Quantify the potential for cumulative positive downstream impacts arising from wetland and other rehabilitation upstream in rivers, enabling this to become a

multiplier of sufficient magnitude to further incentivize investment into brownfields rehabilitation.

- Create the necessary regulatory oversight that measures performance in achieving rehabilitation targets, thereby allowing them to be offset against existing liabilities, rather than expecting a hefty cash deposit as required by NEMA (2014).

The NMLC shown in Figure 2 needs to be better understood, because only by stimulating off-mine jobs can the impact of the RCC be negated. In brownfields sites, this is constrained by the small value of the potential revenue shown in Figure 5. This suggests the need for the quantification of benefits that can be used to offset liabilities that will inevitably fall to the state when the last few mining companies become insolvent by 2019 (Schroder, 2014). In greenfields sites this is not a constraint, so increased attention should be given to stimulating off-mine livelihoods over the life span of the mine concerned. In both of these scenarios the social licence to mine, triggered by the Marikana tragedy, will become increasingly relevant, even if it has not yet been captured by policy or codified into law.

A bite of the reality sandwich – harsh lessons to be learned

The unintended consequences of current environmental and water policy are that the Marikana Massacre[9] has become to the African National Congress in 2012 what the Sharpeville Massacre became for the Nationalist Party government in 1961. Both were watershed events in which policy instability led to a predictable outcome. The handling of both by the state was a determinant of investor confidence going forward. Both defined the capacity of the state to survive under conditions of high risk and inherent social instability. The first triggered an internal uprising based on an armed insurrection, whereas the second has spawned the polarization of the trade union movement, accompanied by the emergence of disruptive politics and vigilante action, leading to a significant loss of investor confidence (Wilson & Cervantes, 2013; Table 1). Policy Paradigm VI arose in 2012 when the events at Marikana revealed the magnitude of the unresolved social tension between the policy paradigm of extraction and the policy paradigm of redistribution (Figure 1). Seen as a subset of the complex whole, water policy plays a small but significant role. Without policy reform, the Gordian Knot inherent in Policy Paradigm VI cannot be undone. Central to this impasse is the increased confusion and complexity with the Department of Environmental Affairs seeming to cut across into the territory of the Department of Mineral Resource and the Department of Water and Sanitation. There are also different regulatory bodies where the radioactivity aspect of uranium is a factor.

The global survey of mining companies (Wilson & Cervantes, 2013) summarized in Table 1 finds that South Africa is currently ranked as one of the worst mining jurisdictions for the categories of Regulatory Duplication and Inconsistency. Significantly, political instability is also listed as a major perceived risk and thus a constraint on investment. If political violence and vigilantism becomes a key element of Policy Paradigm VI, then self-imposed economic sanctions will continue to cripple the economy exactly as externally imposed economic sanctions did during Policy Monopoly II.

Clearly, the global mining industry is not opposed to environmental regulations; jurisdictions like Sweden and Finland rank well in that category. What is needed is coherent policy reform to generate a coherent set of laws and policies that are fair, unambiguous and consistently applied by all regulatory authorities at all levels of the state (national, provincial and municipal). If positive progress is to be made, this reform needs to take full cognizance of the historical context and resultant legacy.

The unintended consequences of South African mining policy, when interpreted through the survey conducted by Wilson and Cervantes (2013), in the context of water resource management and the rehabilitation of mine-impacted aquatic ecosystems are the following:

(1) Current legislation is written with the purpose of ensuring that rehabilitation of greenfield sites is catered to, so as not to repeat the mistakes of Policy Paradigms III and IV. However, when applied to existing legacy brownfield sites, particularly regarding water, it has the unintended consequence of reducing or eliminating the viability of projects by matching current liabilities to historical assets long ago realized and thus unavailable to support these rehabilitation aspirations. This accelerates the collapse of commercial viability, thereby hastening the return of an increased quantum of overall liability to the state as a predictable outcome.

(2) Current policy provides that liabilities created during mining fall to the creator of that liability, so the strategy for the overall rehabilitation remains with the company. This works with greenfield sites where there is no initial liability and the full resource is available to support any liabilities created. However, when applied to legacy brownfield sites, the logical action of executives will necessarily be to dissolve the company prior to the remediation of accrued liabilities, thus shifting the liabilities to the state. In this case, the state needs to provide greater strategic direction in applying its limited assets to support much greater historic liabilities, to assure that these resources are applied in such a manner as not to accelerate the reversion of such liabilities to the state. It must apply such resources in a structured way to maximize the overall impact of the available resources for the benefit of the state, environmental rehabilitation and the restoration of social functionality for impacted communities.

(3) The state should seriously consider amending current policy and legislation to give the regulatory regime the required flexibility to oversee the application of the limited current resources to mitigate the greater historic liabilities. Additionally, in reforming such policy, care should be taken to reduce or eliminate costly and confusing contradictory regulatory regimes.

Conclusion

With the best of intentions, the rigorous application of policy based on international best practices, *inter alia* the polluter-pays principle, does not work when done retrospectively and without consideration of the historic context and appreciation of the unintended consequences. Stated simply, the current suite of policies (water, environmental and financial) is based on the desire to internalize the historic liabilities dictated by the perceived need for state survival during Policy Monopoly II. When applied to Policy Monopoly III this hastens the demise of the mining sector by destroying the business case for investment into brownfield sites. This is exacerbated by the squeeze on profitability arising from the fact that the cost of labour has been decoupled from the productivity of labour through militant union actions (Schroder, 2014). This combination is the reason for the linear collapse shown in

Figure 3. If policy reform distinguishes greenfields logic from brownfields realities in a coherent manner, then this trend can be reversed and capital can be attracted to the rehabilitation of mine-impacted aquatic ecosystems and industrial landscapes. In fact, the value of the gold left in the hundreds of remaining tailings dams is of such a magnitude that recent improvements in metallurgical engineering make it commercially viable to reprocess. But this can be achieved only if new investors are attracted into the industry who are willing to bring old mining companies out of liquidation. This in turn is going to be possible only if a new regulatory regime is capable of quantifying the benefits arising from the consolidation of the many old dumps into a few mega-dumps engineered to twenty-first-century standards, thus allowing that historic liability to be offset by the benefits arising from removal. The capacity of the state, specifically insofar as policy reform and regulatory oversight are concerned, will increasingly be the determining variable if the current trend is to be reversed. South Africa has the potential to succumb once more to festering social unrest if the constraints created by the ruthless application of the externalization-of-costs model inherent in Policy Monopoly II are not effectively countered in a way that incentivizes investment into brownfields operations. Only through evidence-based policy reform can the South African government unravel the Gordian Knot created by the apartheid state in its self-imposed battle for survival during the Cold War.

Water policy thus plays a critical but often invisible role in attracting the level of investment needed to rehabilitate mine-impacted aquatic ecosystems and landscapes. The potential for public–private partnerships as vehicles for rehabilitation is squarely based on the ability of projects arising under such circumstances to attract the level and type of funding needed to make rehabilitation viable. Failing in this crucial area will automatically result in the unintended consequence of nationalizing all remaining environmental liabilities, which will increasingly manifest as a highly impacted aquatic ecosystem and landscape that will pose constraints on future job creation and economic development.

Notes

1. There are four underground mining basins, each hydraulically discrete from the other. The Western Basin has been flooded since 2002 with active decant to surface since that date, whereas the Central and Eastern Basins are in the process of flooding but have not yet reached the decant level where AMD flows uncontrolled into the nearest river. The Far Western Basin is still fully dewatered, but flooding pressures are growing as each mining company ceases to pump, leaving an increased financial burden on the remaining mining companies that will eventually cause their demise.
2. Summarized here from correspondence between the Chamber of Mines and the Department of Water Affairs between 4 November 1975 and 19 January 1976.
3. See http://marikana.mg.co.za/ and http://en.wikipedia.org/wiki/Marikana_killings.
4. The current regulations mean that new investors into old sites are expected to pay a deposit in cash equal to the calculated cumulative liability associated with the site. This is a regulatory issue that could be changed to incentivize investment into such sites with a view to extract the remaining resource to fund rehabilitation.
5. See http://en.wikipedia.org/wiki/Hubbert_peak_theory.
6. See http://en.wikipedia.org/wiki/Association_of_Mineworkers_and_Construction_Union.
7. See http://www.iol.co.za/business/companies/grootvlei-mine-now-stands-in-ruins-1.1075667.
8. See http://www.moneyweb.co.za/moneyweb-mining/blyvooruitzicht-the-fallout-from-a-gold-mines-clos
9. See http://www.enca.com/marikana for more details.

References

Adler, R., Funke, N., Findlater, K., & Turton, A. R. (2007a). *The changing relationship between the government and the mining industry in South Africa: A critical assessment of the far west rand dolomitic water association and the state coordinating technical committee.* Pretoria: Council for Scientific and Industrial Research (CSIR).

Adler, R. A., Claassen, M., Godfrey, L., & Turton, A. R. (2007b). Water, mining and waste: An historical and economic perspective on conflict management in South Africa. *The Economics of Peace and Security Journal*, *2*(2), 32–41. doi:10.15355/epsj.2.2.33

Basson, M. S. (1995). South African water transfer schemes and their impact on the Southern African region. In T. Matiza, S. Craft, & P. Dale (Eds.), *Water resource use in the Zambezi Basin. Proceedings of a Workshop held in Kasane, Botswana, 28 April - 2 May 1993.* Gland, Switzerland: IUCN.

Basson, M. S., Van Niekerk, P. H., & Van Rooyen, J. A. (1997). *Overview of water resources availability and utilization in South Africa.* Pretoria: Department of Water Affairs and Forestry.

Beaufre, A. (1965). *An introduction to strategy.* London: Faber and Faber.

Bega, S. (2012). Removal of mine dump halted. *IOL News*, July 7 2012. Retrieved from http://www.iol.co.za/news/south-africa/gauteng/removal-of-mine-dump-halted-1.1336202#.VH2HQMEaLIU

Blanchon, D. (2001) Les nouveaux enjeux géopolitiques de l'eau en Afrique Australe. *Hérodote Revue de Géographie et de Géopolitique*, Troiseme Trimestre. *102*, 113–137.

Blanchon, D., & Turton, A. R. (2005). Les Transferts Massifs d'Eau en Afrique du Sud. In F. Lasserre (Ed.), *Transferts Massifs d'Eau: Outils de Development ou Instruments de Pouvoir?* (In French) (pp. 247–283). Sainte-Foy, Québec: Presses de l'Université du Québec.

Botha, E. (2013). *Management of Conflict in the West Rand Goldfields due to the Acid Mine Drainage*, (unpublished thesis submitted to the Faculty of Natural and Agricultural Sciences Centre for Environmental Management). University of the Free State, Bloemfontein.

Brodie, B. (1965). *General André Beaufre on strategy: A review of two books.* Santa Monica, CA: The Rand Corporation. Retrieved from http://www.rand.org/pubs/papers/2008/P3157.pdf

Camden-Smith, B., Pretorius, P., Camden-Smith, P., & Tutu, H. (2015). Chemical transformations of metals leaching from gold tailings. Paper accepted for presentation at the 10th International Conference on Acid Rock Drainage (ICARD) and International Mine Water Association (IWMA) Annual Conference. Chile.

Campbell, C. J., & Laherrère, J. H. (1998). The end of cheap oil. *Scientific American*, *278*, 78–83. March. doi:10.1038/scientificamerican0398-78

Coetzee, H. (1995). Radioactivity and the leakage of radioactive waste associated with Witwatersrand gold and uranium mining. In B. J. Merkel, S. Hurst, E. P. Löhnert, & W. Struckmeier (Eds.), *Proceedings uranium mining and hydrogeology 1995.* 1. – 583 S.; Köln (von Loga; ISBN 3-87361-256-9) Freiberg, Germany: *GeoCongress.*

Coetzee, H., Hobbs, P. J., Burgess, J. E., Thomas, A., & Keet, M. (2010). Mine water management in the Witwatersrand Gold fields with special emphasis on acid mine drainage. Report to the Inter-ministerial Committee on Acid Mine Drainage, December 2010, pp 1–128. Retrieved from http://www.dwaf.gov.za/Documents/ACIDReport.pdf

Coetzee, H., Venter, J., & Ntsume, G. (2005). *Contamination of Wetlands by Witwatersrand gold mines – processes and the economic potential of gold in Wetlands. Council for geosciences Report No. 2005-0106.* Pretoria: Council for Geosciences.

Coetzee, H., Wade, P., Ntsume, G., & Jordaan, W. (2002a). *Radioactivity study on sediments in a dam in the Wonderfonteinspruit catchment DWAF Report.* Pretoria: Department of Water Affairs and Forestry.

Coetzee, H., Wade, P., & Winde, F. (2002b). Reliance on existing wetlands for pollution control around the Witwatersrand Gold/Uranium mines in South Africa – are they sufficient? In B. J. Merkel, B. Planer-Friederich, & C. Wolkersdorfer (Eds.), *Uranium in the aquatic environment* (pp. 59–65). Berlin: Springer.

Coetzee, H., Winde, F., & Wade, P. W. (2006). *An assessment of sources, pathways, mechanisms and risks of current and potential future pollution of water and sediments in gold-mining areas of the Wonderfonteinspruit catchment* (WRC Report No. 1214/1/06). Pretoria: Water Research Commission.

Coetzer, P. (2008). Water Management under Pressure: South Africa Might be Facing a Water Contamination Crisis. *Achiever.* 1 March 2008.

COSATO. (2014). Congress of South African Trades Union (COSATO). SA History. Retrieved from http://www.sahistory.org.za/topic/congress-south-african-trade-unions-cosatu

Davenport, J. (2013). *Digging deep: A history of mining in South Africa (1852–2002)*. Johannesburg: Jonathan Ball.

Fawcett, M. H. (1901). *The concentration camps in South Africa*. London: Westminster Gazette.

GDARD. (2011). *Feasibility study on reclamation of mine residue areas for development purposes: Phase II strategy and implementation plan*, December 2011. Report No. 788/06/02/2011 (Final). Umvoto Africa (Chris Hartnady & Andiswa Mlisa) in association with TouchStone Resources (Anthony Turton).

Geldenhuys, D. (1983). The destabilization controversy: An analysis of a high-risk foreign policy option for South Africa. In *Conflict Studies*, No. 148; 11-26. In W. Gutteridge (Ed.), 1995. *South Africa: From apartheid to national unity, 1981–1994* (pp. 42–57). Aldershot, Hants & Brookfield, VT: Dartmouth Publishing.

Geldenhuys, D. (1984). *The diplomacy of isolation: South African foreign policy making*. Johannesburg: Macmillan South Africa.

Grice, A. (1998). Underground mining with backfill. Paper presented to the 2nd Annual Summit of Mine Tailings Disposal Systems, Brisbane, Australia.

Gutteridge, W. (1983). South Africa's national strategy: Implications for regional security. *Conflict Studies*, No. 148; 3–9. Reprinted in In W. Gutteridge (Ed.), 1995. *South Africa: From apartheid to national unity, 1981–1994* (pp. 35–41), Aldershot, Hants & Brookfield, VT: Dartmouth Publishing.

Gutteridge, W. (1984). Mineral resources and national security. *Conflict Studies*, No. 162; 1–25. Reprinted in In W. Gutteridge (Ed.) 1995. *South Africa: From apartheid to national unity, 1981–1994* (pp. 59–83), Aldershot, Hants & Brookfield, VT: Dartmouth Publishing.

Hart, M. H. (2013). *Gold: The race for the world's most seductive metal*. London: Simon & Schuster.

Hartnady, C. J. H. (2009) South Africa's gold production and reserves. *South African Journal of Science, 105*, 328–329. September/October 2009

Heyns, P. (2002). Interbasin transfer of water between SADC countries: A development challenge for the future. In A. R. Turton & R. Henwood (Eds.), *Hydropolitics in the developing world: A southern African perspective* (pp. 157–176). Pretoria: African Water Issues Research Unit (AWIRU.

International Spectator. (2014). *Let's hope against more stupidity in South Africa*. Casey Research. November 2014. Retrieved from http://www.caseyresearch.com/

James, L. H. (1980). Total water strategy needed for the Vaal triangle: Meeting the challenge of the eighties. *Construction in Southern Africa, May*, 103–111.

Jordaan, J. M., Enslin, J. F., Kriel, J. P., Havemann, A. R., Kent, L. E., & Cable, W. H. (1960). *Finale Verslag van die Tussendepartmentele Komitee insake Dolomitiese Mynwater: Verre Wes-Rand, Gerig aan sy Edele die Minister van Waterwese deur die Direkteur van Waterwese*. (In Afrikaans translated as, Final Report of the Interdepartmental Committee on Dolomitic Mine-water: Far West-Rand, Directed at His Excellency the Minister of Water Affairs by the Director of Water Affairs). Pretoria: Department of Water Affairs.

Juwet, G., & Lyssens, M. (2014). *Sculpting the hill, cultivating the valley*. (Thesis submitted for the Master of Science in Urbanism and Strategic Spatial Planning). Catholic University of Leuven, Leuven.

Karis, T., & Carter, G. M. (Eds). (1972). *From protest to challenge: A documentary history of African politics in South Africa 1882–1964*. Stanford: Hoover Institution Press.

Krebs, P. M. (1992). "The last of the gentleman's wars": Women in the Boer War concentration camp controversy. *History Workshop Journal, 33*, 38–56. doi:10.1093/hwj/33.1.38

Liebenberg, B. J. (1987). Botha and Smuts' Rule, 1910–1924. In C. F. J. Muller (Ed.), *Five hundred year: South African history*. Pretoria: Academica.

Liebenberg, I. (1994). Resistance by the SANNC and the ANC, 1912 – 1960. In I. Liebenberg, F. Lortan, B. Nel, & G. Van der Westhuizen (Eds.), *The long march: The story of the struggle for liberation in South Africa* (pp. 8–21). Pretoria: HAUM.

Martin, D., & Johnson, P. (1981). *The struggle for Zimbabwe: The Chimurenga war*. Harare: Zimbabwe Publishing House.

Mbeki, G. (1984). *South Africa: The peasants' revolt*. London: IDAF.

Meredith, M. (2007). *Diamonds, gold and war: The making of South Africa*. Johannesburg: Jonathan Ball.

Morris, D. (1971). War on the Veld. In D. Morris & D. Gardner (Eds.), *The British Empire* (pp. 813–831). London: BBC and Time/Life.

Mudd, G. M. (2007). Global trends in gold mining: Towards quantifying environmental and resource sustainability? *Resources Policy, 32*, 42–56. doi:10.1016/j.resourpol.2007.05.002

Naidoo, S. (2014). Development actors and the issues of acid mine drainage in the Vaal river system. (Unpublished Masters Dissertation in Development Studies). Pretoria: University of South Africa. Retrieved from http://uir.unisa.ac.za/handle/10500/13932

NEMA. (2014). *National environmental management act (Act 107 of 1998)*. Regulations Pertaining to the Financial Provision for the Rehabilitation, Closure and Post Closure of Prospecting, Exploration, Mining or Production Operations. Government Notice 940 dated 31 October 2014. Pretoria: Government Gazette.

Noseweek. (2009, August). Joburg's poisoned well. *Noseweek*, 118, 23–26. Retrieved from http://www.noseweek.co.za/article/2066/Joburgs-poisoned-well

Pakenham, T. (1992). *The Boer war*. London: Harper Perennial.

Pretorius, F. (2001). *Scorched earth*. Cape Town: Human & Rousseau.

Schroder, M. (2014). SA mining: Endgame? Old mutual gold fund manager report to investors in the gold mining industry.

Segar, S. (2013). Here comes the poison, in *Noseweek*, Issue No. 162; 1 April 2013. Retrieved from http://www.noseweek.co.za/article/2934/2013-Here-comes-the-poison

SERI. (2013). Mjadu and Others in Regards the Federation for a Sustainable Environment (FSE) V National Nuclear Regulator and Others (Tudor Shaft). Retrieved from http://www.seri-sa.org/index.php/component/content/article?id=118:mjadu-and-others-in-re-federation-for-a-sustainable-environment-fse-v-national-nuclear-regulator-and-others-tudor-shaft

Tang, D., & Watkins, A. (2011). *Ecologies of gold: The past and future mining landscape of Johannesburg*. Retrieved from http://places.designobserver.com

Toffa, T. (2012). *Mines of gold, mounds of dust*. (Master's thesis for Urbanism and Strategic Planning). Catholic University of Leuven, Belgium.

Turner, J. W. (1998). *Continent Ablaze: The insurgency wars in Africa 1960 to the present*. Johannesburg: Jonathan Ball Publishers.

Turton, A. R. (2000). Precipitation, people, pipelines and power: Towards a political ecology discourse of water in Southern Africa. In P. Stott & S. Sullivan (Eds.), *Political ecology: Science, myth and power* (pp. 132–153). London: Edward Arnold.

Turton, A. R. (2009). South African water and mining policy: A study of strategies for transition management. In D. Huitema & S. Meijerink (Eds.), *Water policy entrepreneurs: A research companion to water transitions around the globe* (pp. 195–214). Netherlands: Edgar Elgar.

Turton, A. R. (2014). Pulling a rabbit from the proverbial hat: Dealing with Johannesburg's slow onset uranium disaster. Forthcoming in the *New South Africa Review* (5). Johannesburg: Wits University Press.

Turton, A. R., Meissner, R., Mampane, P. M., & Seremo, O. (2004). *A Hydropolitical History of South Africa's International River Basins*. Report No. 1220/1/04 to the Water Research Commission. Pretoria: Water Research Commission.

Turton, A. R., Schultz, C., Buckle, H., Kgomongoe, M., Malungani, T., & Drackner, M. (2006). Gold, scorched earth and water: The hydropolitics of Johannesburg. *International Journal of Water Resources Development, 22*(2), 313–335. doi:10.1080/07900620600649827

Van Tonder, D., & Coetzee, H. (2008). *Regional mine closure strategy for the west rand goldfield*. Council for Geosciences Report No. 2008-0175. Pretoria: Department of Minerals and Energy.

Venter, A. J. (1975). *The Zambezi salient: Conflict in Southern Africa*. Greenwich, CT: Devin-Adair Publishers.

Wade, P. W., Woodbourne, S., Morris, W. M., Vos, P., & Jarvis, N. W. (2002). *Tier 1 risk assessment of selected radionuclides in sediments of the Mooi River catchment. WRC Project No. K5/1095*. Pretoria: Water Research Commission.

Wilson, A., & Cervantes, M. (2013). *Survey of mining companies*. Fraser Institute Annual Vancouver, BC: The Fraser Institute. Retrieved from http://www.fraserinstitute.org/uploadedFiles/fraser-ca/Content/research-news/research/publications/mining-survey-2013.pdf

Winde, F. (2006). Inventory of sources and pathways in the catchment. Chapter 3 in Coetzee, H., Winde, F. & Wade, P.W. *An assessment of sources, pathways, mechanisms and risks of current and potential future pollution of water and sediments in gold-mining areas of the Wonderfonteinspruit catchment.* Water Research Commission, WRC Report No. 1214/1/06. pp. 35–53.

Winde, F. (2010). Uranium pollution of the Wonderfontein Spruit, 1997–2008. Part 2: Uranium in water – concentrations, loads and associated risk. *Water SA, 36*(3), 257–278.

Winde, F., & Jacobus Van Der Walt, I. J. (2004). The significance of groundwater-stream interactions and fluctuating stream chemistry on waterborne uranium contamination of streams – a case study from a gold mining site in South Africa. *Journal of Hydrology, 287,* 178–196. doi:10.1016/j.jhydrol.2003.10.004

Wolmarans, E. (2014, February 4). Gold mining's R5bn headache. *The Citizen.* Retrieved from http://citizen.co.za/121189/illegal-minings/

Mine site water-reporting practices, groundwater take and governance frameworks in the Hunter Valley coalfield, Australia

Wendy Timms and Cameron Holley

ABSTRACT

At mine sites in a stressed watershed, groundwater dominated licensed water take, and water-use productivity was dependent on site practices and constraints. Solutions for mining and water in this context include: (1) state-based water governance within a national framework; (2) information tools, including mine site water-reporting frameworks; (3) site water sharing and salt trading; and (4) technologies and leading practices. While water reporting has improved, evaluating the significance of hydrological changes over the long-term remains a challenge, particularly for groundwater and saline discharges to rivers.

Introduction

In the global mining industry water security is recognized as a priority issue for continuity of mining operations, to maintain a social licence to operate with local communities and to gain mine approvals from government regulators (ICCM, 2012). Water security, defined here as an acceptable level of water-related risks to humans, ecosystems and productivity that ensures that there is water available of sufficient quantity and quality (Bakker, 2012), is being undermined by the environmental and social risks of mining. The resulting conflicts and significant delays to mining projects can increase business costs. Franks et al. (2014) found that water resource access and pollution were in the top three causes of conflict for extractive industries, and that community concern and conflict, while possible at any stage of a project, is most likely during the operational phase.

Competition between water users and concerns over reduced environmental flows have become more intense in water-stressed areas of the world. This stress has been acutely felt in parts of Australia (the driest inhabited continent), where a National Plan for Water Security (Australian Government, 2007) has been developed to address over-allocation of water, improve water information, modernize irrigation and support reform of water governance in specific areas.

The Hunter Valley on Australia s eastern coast (Figure 3) has relatively abundant water resources supporting a growing population, highly productive agriculture, strategic regional electricity generation, and a coal mining industry that has expanded significantly in recent years. Although critical water shortages have been mainly restricted to periodic droughts driven by El Niño climatic patterns, the system is fully or over-allocated (NWC, 2010). Whilst basic access rights for water users and the environment have been protected, increasing competition for water appears to have increased the cost of traded water and has continued to raise community concerns in the Hunter Valley.

The management of water allocations and competing demands on water in the Hunter Valley, as in other parts of Australia, involves a hybrid governance system. This system is embodied in the National Water Initiative (NWI), a national agreement between Australia's federal, state and territory governments. The NWI (Australian Government 2006) sets out high-level water management principles (e.g. a water access entitlements and planning framework; water markets and trading; best-practice water pricing; integrated management of water for environmental and other public benefit outcomes), but leaves it to each state and territory government (that have constitutional responsibility over natural resources) to determine how best to undertake their activities. A central pillar of the NWI system is water markets and trading. This involved Australia separating land and water rights, putting a price on water and creating individual water entitlements and allocations that can be traded on water markets. The market is regulated by governments primarily through water plans. Plans set an overarching cap on the available water for consumption ('the consumptive pool'), within a given surface water catchment or aquifer, over a given period (e.g. one year). Individuals' entitlements allow them to be allocated a share of available water from this consumptive pool. People can then choose to buy or sell water, depending on whether their allocation meets their needs during the given period (Australian Government, 2006). In addition to markets and plans, the NWI also establishes a role for state government enforcement of water laws in order to protect water resources and ensure people comply with their entitlements (Hussey & Dovers, 2007). While states, and particularly New South Wales (NSW), have made substantial progress implementing this governance system, a significant 'thorn in the side' of these reforms has been mine water use. Although water use by the mining sector across NSW is very small (1.2% for 2012–03; ABS, 2013), it must be managed on a site and watershed scale.

This weakness arose in part of because of paragraph 34 of the NWI (Australian Government, 2006). Recognizing that mine dewatering and depressurization of coal seams can lead to difficulties in predicting takes and managing impacts, paragraph 34 acknowledged that flexible management arrangements may be required to deal with the specific circumstances of mineral and petroleum sectors (NWC, 2014a). Because of this paragraph, mining remained poorly integrated, and at worse an exception to, the water entitlement and allocation system established under the NWI. However, in recent years, increasing attempts have been made to integrate mining (and coal seam gas) into Australia's water governance framework, in part because of threats to their social licence from community and non-governmental organization's (NGO) protests (Parsons, Lacey, & Moffat, 2014).

This integration is still in its initial stages, and is arguably struggling to keep pace with the fact that energy is now the largest industrial user of water at 15% and growing (Wood Mackenzie, 2013). The 'Troubled Waters' report by these analysts found that expansion of coal mining was in the top three water risks for global energy production. The energy and mining industry is also under increasing scrutiny because water crises, including shortfalls of freshwater, have risen to the third highest global risk (trailing economic and employment concerns), and the highest environmental risk ahead of adaption to climate change (WEF, 2014).

These concerns have led to a range of other non-government and industry led innovations for monitoring and governing the water impacts of mining. For example, water risk for resource companies listed on the Australian Stock Exchange are scrutinized with indicators such as water use normalized to earnings (e.g. ML/million dollars of earnings) (Prior, 2009).

The Global Reporting Initiative (GRI) of sustainability metrics including water withdrawals, discharge and recycling (Table 1) has also become increasingly common in the mining industry (Brown, De Jong, & Levy, 2009; Danoucaras, Woodley, & Moran, 2014). However, critiques of the GRI have emerged, including the possibility that flow volumes may be missed without first reconciling a water account as a basis for calculating indicators or reporting water volumes (Danoucaras et al., 2014). The lack of a water accounting method in the GRI (Morrison & Schulte, 2010) has been acknowledged in the fourth generation of the GRI (2013) guidelines, which have added the requirement that reporting organizations must specify the standards and methods that are applied.

A novel mine water-accounting framework (WAF) has also been developed by the Minerals Council of Australia to enable mine sites to account for, report on and compare site practices in a rigorous, consistent and unambiguous manner (MCA, 2012). The WAF recognizes the vital role of water in mining both as an asset that produces value and as a shared natural resource that requires responsible stewardship

Table 1. GRI and WAF reporting elements.

Reporting	Description
GRI category	
EN8	Total water withdrawn from the environment, categorized by source
EN9	Water sources significantly affected by withdrawals
EN10	Recycling and reuse of water
EN22	Water discharged by quantity and destination
EN26	Identity, size, protected status, and biodiversity value of water bodies and related habitats significantly affected by the reporting organization's discharges of water and runoff
WAF element description	
Contextual Statement	Provides information about the water resources of the region and the catchment in which the sites are located
Input–Output Statement	Lists flows for all input and output categories for the reporting period, along with the change in storage
Statement of Operational Efficiencies	Lists the total flows into the tasks, volume of reused water, reuse efficiency, the volume of recycled water and recycling efficiency
Accuracy Statement	Lists the percentage of flows measured, simulated and estimated
Other reporting elements	
	Volume of water withdrawals permitted by licences
	Quality of water after recycling or reuse
	Water storage capacity
	Regulatory breaches reported
	Future water policies reported

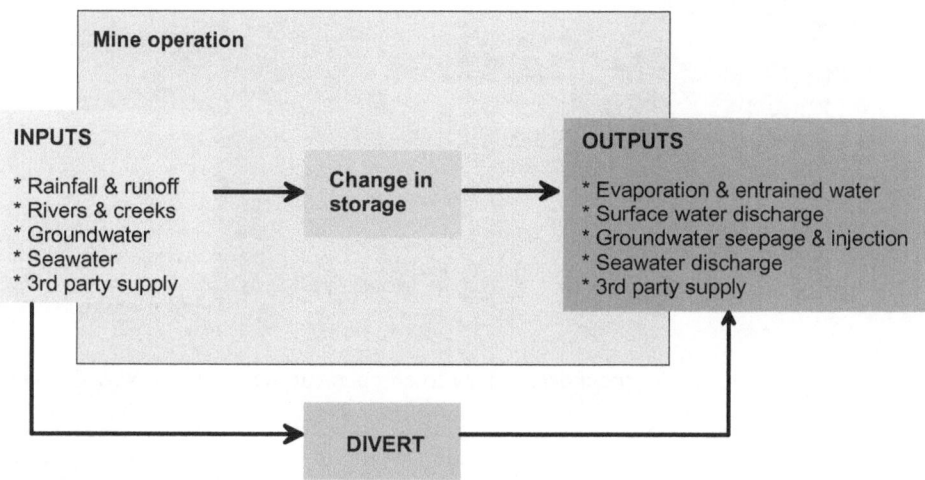

Figure 1. Components of a mine water-accounting framework including key inputs and outputs. Source: after MCA (2012).

(Cote, Cummings, Moran, & Ringwood, 2012). Figure 1 illustrates mine water inputs and outputs from a mine site and some of the key elements of the WAF. Aspects of the GRI and WAF of relevance to this study are further outlined below.

The objective of this paper is to review mine water use (in the Hunter Valley coalfield in Australia), and to evaluate the growing range of potential governance tools and solutions to ensure that sufficient water of a suitable quality is available for users. In particular, it highlights the recent adoption of mine water-reporting frameworks that have enabled information to be compiled and evaluated at a watershed scale, and the development of new indicators such as the water-use productivity curve for the mining industry. It is demonstrated that these reporting frameworks, along with three other tools and governance approaches, form a suite of possible solutions to improve security of water supplies to all water users in mining areas such as the Hunter Valley. These potential solutions include: (1) state-based water governance within a national framework, (2) information tools such as mine site water-reporting frameworks, (3) site water sharing and salt trading, and (4) technologies and leading practices.

Mine water issues at a watershed and site level

Mine water issues may be perceived as problems that block mining development, as challenges to be transformed, or somewhere between these views. Increasing community and regulatory scrutiny of water management by the mining industry requires evidence-based evaluation and articulation of actual, rather than perceived, issues. This paper focuses on mine water at a critical scale between a watershed and a mine site level, rather than across vast regions, within jurisdictional boundaries or at a company level. Several challenging issues are documented on the basis of recent mine water data, with a particular focus on groundwater.

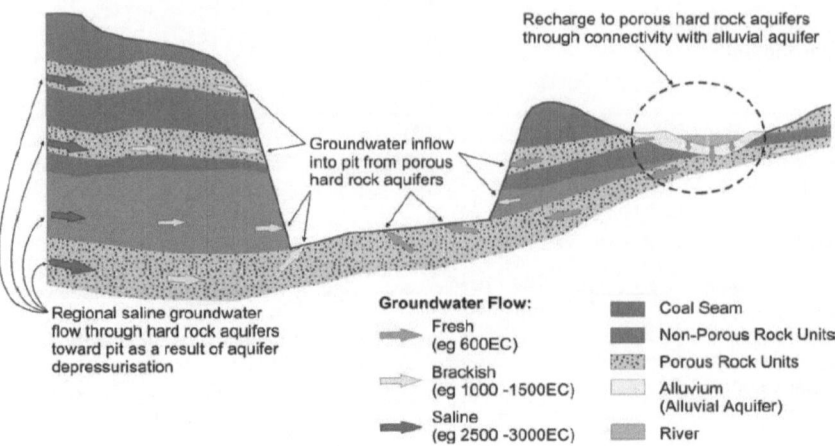

Figure 2. Potential pathways for groundwater flow to an open-cut mine. Source: NSW Government (2005).

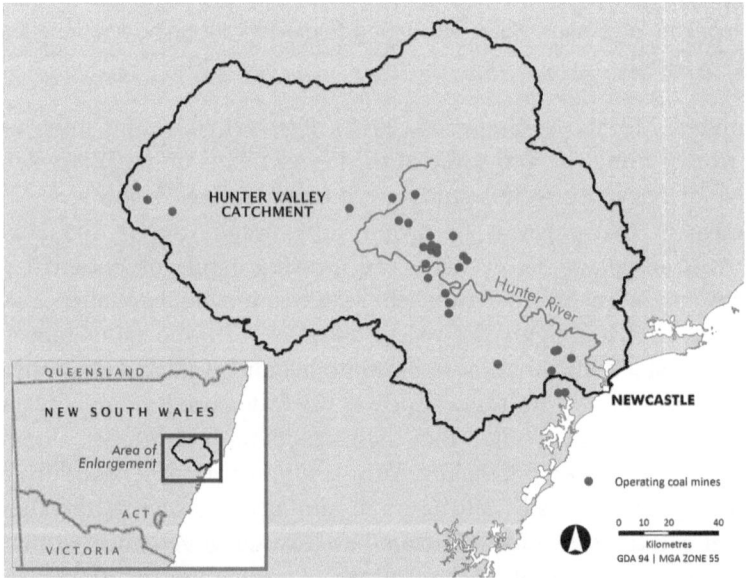

Figure 3. Hunter Valley watershed showing operating mine sites. The inset shows the study area within the state of New South Wales, Australia.

The volume of groundwater flow to open pit or underground mine workings can have implications for safe and efficient extraction and for potential impacts on shallow aquifers, or nearby rivers, lakes and wetlands. Figure 2 shows potential groundwater flows to an open pit from permeable rock aquifers, and from alluvial aquifers adjacent to a river (NSW Government, 2005).

Flows to underground workings can occur via geological structures, flow from old mine workings, from within the worked coal seam and from subsidence induced fractures. Tunnelling and coal mining beneath large water bodies occurs where geological conditions provide natural barriers to groundwater flow. Fractures and fault zones can be conduits or barriers to flow, with permeability typically reduced by infill and stress regimes in some areas, particularly with the increasing depth of cover of recent mining operations in the Sydney Basin (Tonkin & Timms, 2015).

Study area and methods

The Hunter watershed in New South Wales

The Hunter River watershed (approximately 22,000 km^2) is located on the NSW coast north of Sydney (Figure 3). Average annual rainfall is approximately 640 mm at Muswellbrook. The Hunter River flows for 460 km from headwaters to the West of the Barrington Tops (over 1000 m elevation), southward and then eastward to the Pacific Ocean. The Goulburn River, a major tributary of the Hunter River, flows from the Western part of the watershed.

Mining and agriculture support populations in towns that are found mainly along river valleys along with electricity generation and tourism that is particularly associated with the wine industry. Cattle, pasture production, horse breeding and viticulture dominate agricultural activities with the town of Scone and its 70 horse studs, being one of the equine capitals of the world.

Watershed management priorities are targeted in the Hunter–Central Rivers Catchment Management Plan (now changing under Local Land Services reforms), and water-sharing plans (WSPs) were developed under the Water Management Act 2000 (discussed below). The Hunter Regulated River, from Glenbawn Dam (750 GL), located east of Scone, downstream to the estuarine limit, is managed under a WSP that specifically accounts for flows that are regulated by dam releases. Glenbawn Dam and Glennies Creek Dam (283 GL) are both located in the Upper Hunter watershed. The watershed is fully allocated (Dragun, 1985), in that an embargo on applications for new commercial water licences has been in place for the Hunter Regulated River since 1982.

Mine sites and methodology

Data were compiled from publically available reports by mining companies and government water regulators. A preliminary evaluation of reporting practices at all 24 currently operating mines in the region was undertaken. Ten mine sites were then selected for more detailed analysis of water inputs, outputs and use (Table 2). Selection criteria included availability of relatively comprehensive site water information, representing different types of mining operations (underground, open cut and combined underground and open-cut operations) that were geographically spread around the Hunter valley.

The collection of data was extracted from reports available from mining companies. Over 50 publically accessible annual environmental management reports (AEMRs) and complementary reports including site water management plans (SWMPs) contributed to this study.

Table 2. Mine site information for selected mines in the Hunter Valley coalfield, Australia.

Mine	Mine type	Coal type	Product coal, 2012 (Mt)
U-2	Underground	Coking	1.2
OU-5	Open cut and underground	Thermal	3.0
O-9	Open cut	Thermal and coking	12.0
O-13	Open cut	Thermal	8.2
OU-16	Open cut and underground	Thermal and coking	10.9
O-17	Open cut	Thermal	1.0
O-18	Open cut	Thermal	1.8
U-19	Underground	Coking	1.5
U-24	Underground	Thermal	3.3
OU-25	Open cut and underground	Thermal	5.3
O-26	Open cut	Thermal	5.0

Note: Sites 17 and 18 were recently combined for reporting purposes.

Solutions for mine water management

Based on our analysis of data from the 10 mines sites, the applicable laws and the literature, we identified several potential solutions to mine-site water challenges in a regional context including: (1) state-based water governance within a national framework, (2) mine site-scale water reporting, (3) water sharing between mines sites, and (4) technology and leading practice for reducing water use and improving quality of discharge.

State-based water governance solutions within a national framework

As discussed above, serious attempts have been made to integrate mines into the existing NWI water framework of markets and WSPs and regulation. The integration of mining into Australia's water framework commenced with calls by national bodies to ensure that the mineral industry's impacts on water resources, including cumulative impacts, were explicitly addressed (NWC, 2014a). Although there is a legacy of excluded mining sites, and some states continue to exclude mining, NSW now requires minerals operations to hold a volumetric licence (i.e. an entitlement) for any water taken (although prospecting and fossicking remain exempt) (NWC, 2014b). As many water systems can be fully allocated, licences will often be obtained by mines through the market and trades under water plans (known as WSPs in NSW) (NSW Department of Primary Industries (NSW DPI), 2012; NWC, 2014b; s 60I Water Management Act 2000; Wheeler, Loch, Zuo, & Bjornlund, 2014) or other means where water sources are outside planned areas (e.g. Water Act 1912).

In the Hunter Valley region there are two WSPs that apply at a local level for specific water sources, namely the Hunter Regulated River (commenced 2004), and the Hunter Unregulated and Alluvial Sources (commenced in 2009). A separate WSP for Wybong Creek is to be included within the later. There are currently no WSP for porous and fractured rock aquifers in this area, partly because allocations for mining operations that draw water from these sources are considered to be significantly smaller than the estimated annual average recharge.

Water access licences and shares for a surface water source (the Hunter River regulated source) are summarized in Table 3 as an example of this form of water governance. Water is shared out between access licences based on available water

Table 3. Water access licences for the Hunter River Regulated Source and shares at the start of the WSP.

Access licence category	Units	Total share component	%
Major utility	ML/year	36,000	74
Local water utility		10,832	49
Domestic and stock		1738	1
	Total ML/year	48,570	
High security	Unit shares	22,159	11
General security		128,163	64
Supplementary water		49,000	25
	Total unit shares	199,322	

Source: NSW Department of Infrastructure Planning and Natural Resources (2004).

determinations (AWDs) that are made at the start of each water year for the different categories of access licence. For 'high priority' access licences – domestic and stock, local water utility and major water utility – the AWD will provide allocations equal to 100% of the share component volume in all but the most exceptional drought years. High security-regulated river access licences will receive at least 0.75 ML/unit share in all but the worst drought years. High security users receive 1 ML/unit share in years when general security access licences have been allocated at least 0.5 ML/unit share.

Water resources in some areas of the Hunter Valley are over allocated and would not be sustainable if actual water use were equal to allocations (e.g. Pages River, Dart Brook and Wybong Creek in the Upper Hunter). In reality, the sustainability status of a water source depends in part on water licences held that are partially used or not used (i.e. 'sleeper' licences). In other words, physical water scarcity may not be critical even in a watershed where water resources are fully allocated. Dragun (1985) recognized that water markets and effective governance could alleviate emerging physical water scarcity in the Hunter Valley.

The status of the Pages River and potential impacts of mining was a key factor in the proposed Bickham open-cut mine not gaining regulatory approval (NSW Government, 2010). An important note is that high-security water licences may be held by some mining sites because the flows are constant and may not be possible to eliminate (e.g. incidental seepage from river alluvium to an open-cut pit). Given the nature of these flows, high-security water licences are a necessary and appropriate form of governance.

Despite this, the effectiveness of this overall governance regime has been limited by a number of broader implementation challenges, including delivering effective and comprehensive compliance and enforcement (Holley & Sinclair, 2012) and gaps in scientific understanding about the potential water-related impacts of the minerals industry.

Both the federal and NSW governments have undertaken reforms to try to address these weaknesses. In terms of the federal government, these reforms have included the National Framework for Compliance and Enforcement Systems for Water Resource Management, the National Partnership Agreement on Coal Seam Gas and Large Coal Mining Development, bioregional assessments and the establishment of the Independent Expert Scientific Committee to provide advice to governments and regulators on projects where there is likely to be a significant impact on water resources. The Commonwealth also introduced the 'water trigger' into the Environment Protection and Biodiversity Conservation Act 1999 (Cth) (EPBCA). Where a proposed large coal-mining development could have a significant impact on a water resource, the commonwealth government assesses and imposes conditions on these projects. The water trigger was developed partly to overcome any perceived or actual conflict of interest of states' assessing developments (as they receive royalty payments from mining developments). However, the trigger is currently subject to an independent review and proposed amendments to the EPBCA aim to reduce the commonwealth's role in this process, by devolving responsibility to the states and territories to undertake the assessment and approval process (McGrath, 2014; NWC, 2014a).

At the NSW level, a new business model and the hiring of new staff have been undertaken by the regulator to enhance compliance and monitoring capabilities. At the time of writing, the government had also introduced a NSW Gas Plan, Codes of Practice (e.g. well integrity), new exploration regulations and an Aquifer Interference Policy (NSWDPI 2012). The latter sets out a system of triggers for water tables and groundwater pressures such that where triggers were exceeded, additional studies, contingency plans and 'make good' arrangements are required (NWC 2014b). The interference policy is supported by a Strategic Regional Land Use Policy and Plans (such as the Strategic Regional Land Use Plan – Upper Hunter) which attempts to better plan and resolve conflicts between mining and agriculture (and their nexus with water). This policy introduced a new 'Gateway Panel' whose primary responsibility is to conduct a pre-assessment of the agricultural impacts of state significant mining or gas development proposals on various types of strategic agricultural land (i.e. highly productive with unique natural resource characteristics). While this was an important attempt to better identify/ address trade-offs, this approach has been criticized for significant gaps in knowledge relating to water systems/cumulative impacts (Owens, 2012). In addition, the gateway panel cannot refuse a development – it can only issue a certificate or a certificate with conditions to be addressed during the following development application process (see the Environmental Planning and Assessment Act 1979 (NSW); Holley & Kennedy, 2015).

Under this subsequent development application process, the decision maker must consider the economic, social and environmental impacts of the mine (New South Wales Government, 2007; Kennedy 2016; Holley & Kennedy 2015).

In summary, despite the growing integration of mining into Australia's water governance framework, many of the recent reforms remain at an early stage and there is still a long way to go to achieve successful implementation. This suggests there is significant merit in looking to other possible complementary solutions to managing mine site water. These are discussed below.

Information solutions

The mining industry has developed and adopted water-reporting frameworks ahead of some other sectors, despite often using a relatively small proportion of total water supply. The five relevant GRI reporting categories (GRI, 2013) and components of the WAF (MCA, 2012) are outlined in Table 1.

An explanation of WAF terms relevant to this study, in relation to water governance terminology in NSW, includes the following: tasks, inputs, outputs and diversions.

Mine water use was defined by WAF as a task, e.g. dust suppression, product coal, coal reject and water in tailings, rejects and product coal. WAF defines an input as a volume of water that is received by the operational facility for intended use by the operational facility. Inputs include water sources such as groundwater, surface water, harvesting of rainfall runoff and water supplied by a third party. Outputs are flows from the operational facility to these water sources. Inputs exclude diversions that are flows directly from inputs to outputs without being used.

Groundwater could be sourced from a bore supply, from active dewatering, or is received as seepage to a mine in excess of water-use needs. In NSW all groundwater extraction, withdrawal or seepage is regarded as a water take that requires licensing under the state-based governance framework (discussed above), as does extracting water from a river or lake. Form a governance perspective, it is important to note that water use may not exceed a water access licence, a volumetric allocation that is determined to be sustainable for the watershed.

Water is regarded by the WAF to have one of three statuses: raw, worked and treated worked. Raw water is water taken from surrounding systems, including the environment. Reused water is the worked water used for tasks on the site, and distinguished from recycled water is worked water that has been treated before use.

Current mine development consents in NSW require reporting on all GRI indicators, and some, but not all, the information required by WAF (Leong, Hazelton, Taplin, Timms, & Laurence, 2014). These reporting requirements may not, however, require the detail that formal GRI and WAF demand. Beyond NSW, it is recognized that mine site-level information may be problematic due to inconsistent terminology and reporting formats (Cote et al., 2012). Furthermore, use of benchmarked or equivalent quantities of extractions in different locations may have significantly different hydrological effects (Irbaris, 2009; Leong et al., 2014). General-purpose water accounts could provide a watershed context for mine site-water accounts. However, under Australian Water Reporting Standard 1, these are currently only available for NSW watersheds that are within the Murray–Darling Basin (NSW Office of Water, 2014).

Mine site water-reporting practices in the Hunter Valley coalfield

A first-pass evaluation of reports by all current coal mining operations in the Hunter Valley coalfield found that 96% of mine sites ($n = 24$) provided information outlined by the GRI requirements, although not with a consistent degree of detailed information (4% were missing such information). In addition, 92% of these mine sites provided information as required by the WAF reports, even though the framework is relatively new. Similarly to GRI, the completeness of this reporting varied and 8% of sites were missing relevant WAF information.

The accuracy or adequacy of the reports from a technical point of view was not considered, simply whether or no information on mine water had been reported or not. Mine sites with excellent water information availability (including most components required for GRI and WAF) were Hunter Valley Operations and Mt Thorley Warkworth sites. The majority of mine-site water reports were not formal input–output statements fully consistent with WAF, but were focused on describing the flows of water to operational tasks. Nevertheless, for most sites, information required to generate an approximate input–output statement of water takes and discharges was possible to identify in AEMRs, combined with complementary reporting such as SWMPs.

The source of water take was found to vary from exclusively groundwater to dominantly surface water at the 10 mine sites (Table 2). Groundwater was the largest licensed water take from the environment for most of the sites, yet was on average less than half the allocations that these mines were licensed to access. However, groundwater flows are inherently difficult to quantify due to natural subsurface variability, and the WAF requires differentiation of groundwater as inputs or diversions. Inputs such as seepage to a pit that is subsequently used for a task such as dust suppression would be accounted for as an input, whereas that seepage that is directly discharged to surface water or re-injected into a different aquifer without being used would be accounted as a diversion. The inherent difficulty of measuring, modelling or estimating groundwater flows, except where extractions are piped and metered, also means that volumes are typically less accurate than other flows defined in WAF.

Many of the mine sites in this evaluation introduced dedicated environment and water reporting in the early 2000s, and other sites in the late 2000s. It is apparent that older development consent conditions may not have included rigorous disclosure requirements. These study findings were consistent with Leong et al. (2014) who found that there are now multiple reporting requirements for mine water in NSW.

Water sharing and trading solutions

Water security in the Hunter Valley is improved by sharing water by transfer agreement, by permanent or temporary water transfers subject to licensing and regulatory rules, and by trading of salt discharge permits. Loss of agricultural production or opportunities due to demand for water by coal mining must be offset by economic benefits to NSW and the Hunter Valley, and to compensation to affected landholders for either permanent or temporary losses incurred (NSW Government, 2005). Compensation to landholders affected by mining operations is payable according to project approval conditions, or under the Mining Act 1992. Conditions for mining may include, for example, making good provisions to provide a landholder with alternative water supplies should water levels fall below an agreed trigger level. Mining operations must share water through formal transfer arrangements or seek water licences on the market, given that all water resources in the Hunter Valley are fully allocated and no new licences are available.

Water sharing between mine sites

Surplus water at mines is used to supplement water requirements at nearby mines with a water deficit. The Ulan underground mine has been transferring water by pipeline to the Moolarben open-cut mine (Figure 3, western Hunter Valley), which is located up the dip of the coal strata and further from water sources. Ulan coal has transferred from 220 to 701 ML each year (from 2011 to 2013) of mine water to the Moolarben mine (Ulan Coal Mines Ltd, 2013). In 2012, the 220 ML transfer was approximately 9% of total water use (including evaporation), or approximately 4% of groundwater seepage into the mine. In 2012, Moolarben reported 613 ML transferred (Moolarben Coal,

2013), which accounted for approximately 39% of total water take, or 45% of total water use. Discrepancies in reported water transfer are attributed to different start dates for the annual reporting periods and annual flows that are estimated based on less than 12 months of data, issues commonly observed in water reports.

Hydrological modelling by David and Dundon (2010) highlighted future opportunities to expand the water-sharing scheme between mines that would alleviate the need for additional groundwater pumping at Moolarben and Wilpinjong mines, and reduce the need for water treatment and offsite disposal of surplus water at Ulan. Based on the current mine plans and schedules of the three mine projects, and mine water inflow rates that were modelled by David and Dundon, the surplus water produced by dewatering at Ulan would be sufficient to offset the shortfalls at two other sites.

Opportunities for water sharing between mine sites will likely emerge elsewhere, as a form of water trading, driven by lack of new water sources and a need to reuse mine water. Although the cost of pumping and piping infrastructure for water transfer schemes would typically exceed water sourced at or adjacent to a mine site, market prices for water trades could further drive such schemes. Mine-site water transfers involve multi-tiered governance arrangements involving licensing requirements, market trading and water planning discussed above (Hussey & Dovers, 2007).

Hunter Valley salt-trading scheme (HVSTS)

The HVSTS is a unique solution to the problem of saline water generated by mining. As in many areas of Australia, salinity is a primary water quality concern, with mining and agricultural activities exacerbating salt fluxes in landscapes that are naturally saline.

Farmers, concerned that increasing river salinity was impacting the irrigation of food and fodder crops, worked together with the mining and power-generation industries to develop a world-leading salt-trading scheme (Krogh, Dorani, Foulsham, McSorley, & Hoey, 2013). Discharge of saline water from mine sites to the Hunter River has been prohibited since 1994, except as licensed discharges during high-flow events, subject to the rules of the HVSTS. After a successful pilot phase, the scheme was formalized in 2002 when the Protection of the Environment Operations (Hunter River Salinity Trading Scheme) Regulation 2002 commenced. A 10-year review of the scheme is currently in progress.

Thirteen mines and three power stations currently participate in the HVSTS, trading salt credits permitting controlled release of saline water into the river during high-flow events (Vink, Hoey, Robbins, & Roux, 2013). Discharge volumes, salt concentrations and dilution factors are continuously monitored to achieve salinity targets at key points along the river. For example, allowable maximum salinity in the river during discharge events is set as 900 μS/cm at the most downstream monitoring point of the scheme.

A recent review of salinity trends (Krogh et al., 2013) since the 1970s found no evidence of increased groundwater salinity or river water salinity; however, the review was limited by a lack of recent data from government-monitoring bores to compare with historical records. Cyclical changes in water quality due to many factors such as rainfall were evident. Overall, Krogh et al. (2013) concluded that the scheme has improved river water salinity in the mid to lower catchment, a success given the very low flow rates during the 1990s and 2000s. The effectiveness of the scheme in other areas of the catchment was less clear, with no significant trend in water salinity in the upper catchment despite several salinity hotspots or periods of time when salinity values exceeded fresh water quality guidelines.

Mine sites were estimated to have contributed 10–20% of the salt load in the Hunter River by controlled discharges under the HVSTS, with higher values in recent years (Krogh et al., 2013). However, only 25–50% of the total available discharge (TAD) opportunities were utilized by scheme participants, although utilization has improved over time. The mining industry has called for further improvements in upstream monitoring and predictions that would provide more lead-in time to prepare to discharge, along with a number of initiatives to increase discharge opportunities (NSW MC, 2014).

Whilst the mining industry acknowledges the success of the scheme, a number of potential improvements have been identified that would improve environmental outcomes. The NSW MC (2014) outlined a number of priorities including a need for studies to alter salinity targets, the management of other salinity sources particularly from the Goulburn area, examining the dilution of other pollutants such as metals, and allowing discharge during floods and if the discharge is less saline than the river.

Given the increased attention to mine water reporting, there is an opportunity to examine temporal and spatial patterns of flow volume and contaminant loads that occur near mine sites. Mining industry support for increased public access to water information (NSW MC, 2014) could include compilation of site-based groundwater data to augment the lack of groundwater data that were identified by Krogh et al. (2013).

Technology and leading practice solutions

Mining at all stages (including exploration, planning, permitting, operation, closure and rehabilitation) requires water technology to be reliable in all aspects (Brown, 2010). Technologies can improve mine water productivity where and when there is water scarcity, and improve discharge quality if there is a water surplus.

Water productivity in mining operations

Water-use productivity (229–2075 L/tonne product, average 771, $n = 10$; Table 4) was dependent on site practices and constraints. These productivity values included water

Table 4. Indicators of water take and use litres/tonne (L/t) coal product, 2012.

| Site | Water take | Water use | | |
	Groundwater and river water	(1) Including evaporation and entrainment	(2) Excluding evaporation and entrainment	Difference (1) – (2)
U-2	332	632	632	0
OU-5	1027	1189	245	944
O-9	437	678	514	163
O-13	151	593	496	96
OU-16	321	1041	986	55
O-17	1930	2075	2059	17
U-19	430	430	0	430
U-24	376	229	227	2
OU-25	1115	467	290	177
OU-26	82	378	347	32
Average	620	771	580	192

lost by evaporation, and entrainment of water in the product and waste. By excluding evaporation and entrainment losses, a lower average water-use productivity value of 580 L/tonne was estimated for these mine sites. In areas of water shortage, technologies to decrease evaporation losses and recover water from waste could thus be beneficial.

By comparison, 402 L/tonne of product coal was reported by Kunz and Moran (2014). It is not clear, based on the available information, whether the higher average value found in this study can be attributed to differing geographical conditions, terminology and definition or other factors.

A water-use productivity curve was developed for mining operations that is analogous to the cost curve for mining operations (Figure 4). For this indicator, water-use productivity was defined in terms of good produced per unit water (Gleick, Christian-Smith, & Cooley, 2011). As with any cost curve for production, this new tool can be readily adopted by the minerals industry to gauge its comparative position, specifically in water use/tonne of product. This cost curve could complement established indicators where water use is normalized to earnings and other factors (Prior, 2009; Wood Mackenzie, 2013).

Figure 4 shows that high water take and use is not directly related to the type of mine (underground, open cut) or total tonnes of production at a site, suggesting that site-specific practices, technologies and constraints are critical to improving how efficiently water is used. Maximizing water productivity, however, is not a 'silver bullet' for all water problems, and mainly suits arid rainfed areas (Amarasinghe & Smakhtin 2014) or, in this study, watersheds that are fully allocated even if physical water scarcity is not critical.

New technologies and leading practice

As new challenges emerge and new solutions are developed, it is important that leading practice be flexible in developing solutions that match site-specific requirements

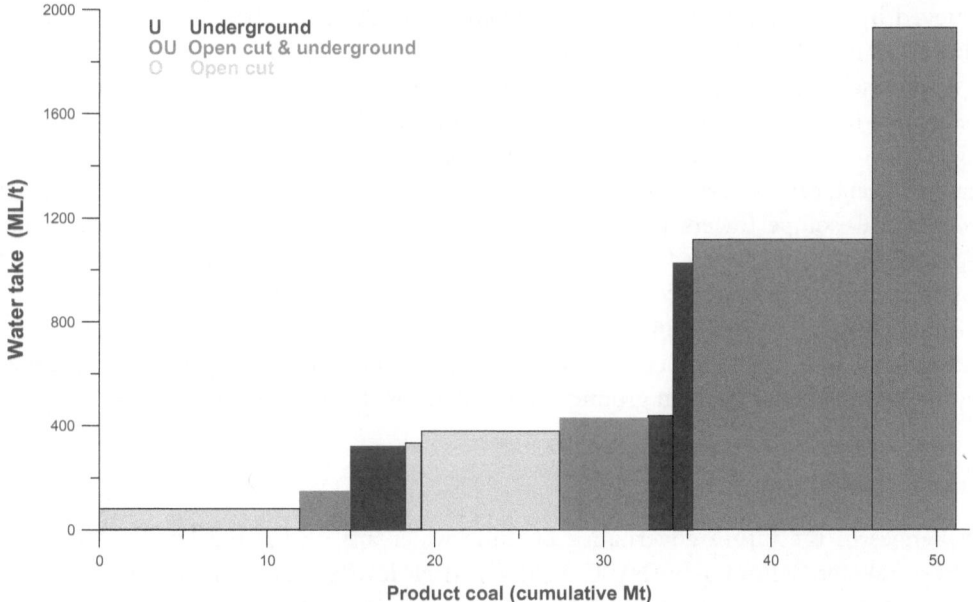

Figure 4. Water productivity curve comparing water take (ML/t of surface and groundwater) and cumulative coal production (Mt).

(Laurence, 2011). A key point of leading practices is that multidisciplinary approaches that combine a number of independent and complementary assessments would increase confidence in the findings. A number of examples of leading practices in water management at mine sites are detailed as part of a series of guidebooks developed with the mining industry (Australian Government, 2008).

As leading practice is as much about possible approaches as it is about a fixed set of practices or a particular technology, regular updates on developments are warranted. One such development is significantly increased treatment of seawater and brackish or saline groundwater for mine site use, or to meet water quality criteria for discharge (Bluefield Research, 2014). This analysis found that over US$15 billion-worth of desalination plants and supply pipelines dedicated to mining are in development around the world. Examples of specific leading practices focused on improved techniques to assess vertical connectivity of groundwater through strata overlying underground mining were outlined by Timms, Acworth, Hartland, and Laurence (2012).

Practices for constructed flow barriers between open-cut mines and surface waters are developing with each new system that is designed. For example, a low permeability barrier (LPB) wall between an open pit and a river is typically constructed of soil–bentonite with a monitoring bore nest on either side of the wall to assess the integrity of the LPB. LPBs have been constructed since the 1980s to allow open-cut mines to be developed adjacent to the Hunter River (Timms, Liu, & Laurence, 2013). Further work is required on the long-term management, breaching or removal of these barriers after mine rehabilitation is completed.

Recent mining project approvals in the Hunter Valley have included consent conditions for LPBs where palaeochannels could provide flow connections between an open pit and a river. Construction cost is a significant concern and largely site specific, with an LPB that was 15 m depth over a length of 1200 m constructed at a cost of A$3 million (Daracon, 2010). Significant cost savings of up to A$1.1 million may be achieved by reducing bentonite content from 8% to 2% assuming LPB dimensions of 1 km × 30 m × 1 m (Bouzalakos, Crane, Liu, & Timms, 2014).

Additional targeted hydrological monitoring at open-cut and underground mines that are located in areas of high water stress are recommended to identify water savings and efficiencies. The significance of mining-induced seepage near the surface and within shallow strata can be evaluated with a complementary suite of techniques including geochemical-isotope tracers (David, Timms, & Baker, 2014), aquifer interference tests and hydraulic tomography (Illman, 2014), and quantitative evaluations of the surface water–aquifer interaction (Rau, Andersen, McCallum, Roshan, & Acworth, 2014). Advanced numerical modelling may also be appropriate in some cases, if suitable data are available, to understand better coupled geo-mechanical and hydrogeological processes and to quantify better flows in groundwater systems with complex architecture.

Future challenges

Challenges in the future governance of mine water are recognized by the Australian National Water Commission (NWC) (2014b). High-level priorities include developing a coordinated jurisdictional approach between the mining industry and multiple government agencies. Other priorities include enabling accurate accounting of water take and implementing adaptive management practices. The NWC has highlighted that increased adoption of consistent WAFs by the mining industry with a view to future requirements will generate benefits including a reduced regulatory burden. It is recognized that all water users will benefit from more effective conditions and increased scientific capacity and groundwater understanding.

Several technical aspects of mine water management require further attention including saline water in mine voids and the effects of climate variability on mine water, both over a decadal time scale and longer. Such issues are relevant at many mine sites around the world and the application of the new aquifer interference policy in NSW has broader implications. The possibility of dewatering interactions and storage changes with complex aquifer–aquitard systems that are not readily measurable means that advanced techniques in the research domain need to be applied in water management, that is, as a leading practice (Timms et al., 2012).

Further work is also required to extend this study across all mines in this watershed, and watersheds elsewhere, evaluating changes over time as the WAF is increasingly adopted as a reporting standard in the mining industry to complement GRI reporting. It is recommended that additional opportunities for water sharing between multiple users are explored. There are opportunities to reduce evaporative losses to improve water productivity/tonne of coal and to reduce the water risk to mining operations, agricultural water users and environmental flows.

Conclusions

Improvements in water reporting have enabled mine water issues at a watershed scale to be evaluated in unprecedented detail and a range of solutions were highlighted to improve the security of water supplies for all water users. These solutions included water-governance frameworks, information tools such as mine-site water reporting, water sharing and trading, and technologies and leading practices. Each solution has something valuable to offer, and none is 'right' or 'wrong' in the abstract (Gunningham & Holley, 2010). There is no 'silver bullet' for water-management challenges (Gleick et al., 2011). Rather, each solution makes a differing (and sometimes overlapping) contribution. This paper's goal was to highlight these major solutions relating to mine sites in a region with increasing competition for water, so as to lay bare some of the key innovations for both Australia and other countries. While the feasibility and effectiveness of these solutions is likely to vary depending upon context (e.g. the type and size of mining sites and the policy and development context of a given country), there are also a number of insights that have wider application. In particular, the paper has offered novel insights by compiling dominant water inputs (takes) and outputs (discharges) from mine sites for the first time across a watershed. Sustainability indicators for water take relative to licensed allocations combine elements of water accounts with water-governance arrangements, and thus serve to strengthen reporting on the significance of water withdrawals. A new water-use productivity curve was developed, analogous to a mining cost curve, showing that high water take and use is unrelated to the type of mine or mine-site coal production, suggesting that site-specific practices and constraints are critical to improving how efficiently water is used. While water reporting has improved, evaluating the significance of hydrological changes over the long-term remains a challenge, particularly for groundwater and saline discharges to rivers.

Acknowledgements

Professor Andy Baker of the UNSW Connected Waters Initiative Research Centre and Associate Professor David Laurence of the Australian Centre for Sustainable Mining Practices are acknowledged for inspiring and supporting this research. Cristiane Dias de Novaes is thanked for her assistance in the early stages of this research.

Disclosure statement

The authors state there is no financial interest or benefit they have arising from the direct applications of this research. This independent research did not receive financial support from the mining industry or mining stakeholders.

Funding

Cameron Holley acknowledges funding from an Australian Research Council Discovery Early Career Researcher Award [grant number DE140101216] that supported his research for this article.

References

ABS. (2013). Water account, Australia, 2012–2013. Retrieved from http://www.abs.gov.au/aus stats/abs@.nsf/0/34D00D44C3DFB51CCA2568A900143BDE?Open

Amarasinghe, U., & Smakhtin, V. (2014). Water productivity and water footprint: Misguided concepts or useful tools in water management and policy? *Water Productivity and Water Footprint: Misguided Concepts or Useful Tools in Water Management and Policy? 39*(7), 1000–1017.

Australian Government. (2006). *Intergovernmental agreement on a National Water Initiative.* Canberra, Australia: Commonwealth of Australia and the Governments of New South Wales, Victoria, Queensland, South Australia, Western Australia, Tasmania and the Australian Capital Territory and the Northern Territory.

Australian Government. (2007). A national plan for water security. Retrieved from http://nailsma.org.au/sites/default/files/publications/national_plan_water_security2007.pdf

Australian Government. (2008). Water management. *Leading practice sustainable development program for the mining industry.* Retrieved from http://www.industry.gov.au/resource/Documents/LPSDP/LPSDP-WaterHandbook.pdf

Bakker, K. (2012). Water security: Research challenges and opportunities. *Science, 337*(6097), 914–915. doi:10.1126/science.1226337

Bluefield Research. (2014). Water solutions for global mining: Competition, costs & demand outlook, 2014–2019. *Bluefield's advanced water treatment & desalination insight service.* Retrieved from http://bluefieldresearch.com/advanced-water-treatment-and-desalination/

Bouzalakos, S., Crane, R., Liu, H., & Timms, W. (2014, May). *Geotechnical and modelling studies of low permeability barriers to limit subsurface mine water seepage.* Paper presented at the 4th international conference on Water Management in Mining, Vina del Mar, Chile.

Brown, A. (2010). Reliable mine water technology. *Mine Water and the Environment, 29*, 85–91. doi:10.1007/s10230-010-0111-7

Brown, H., De Jong, M., & Levy, D. (2009). Building institutions based on information disclosure: Lessons from GRI's sustainability reporting. *Journal of Cleaner Production, 17*, 571–580. doi:10.1016/j.jclepro.2008.12.009

Cote, C. M., Cummings, J., Moran, C. J., & Ringwood, K. (2012). Water accounting in mining and minerals processing. In J. M. Godfrey & K. Chalmers (Eds.), *Water accounting – International approaches to policy and decision-making* (pp. 91–105). Cheltenham: Edward Elgar.

Danoucaras, A. N., Woodley, A. P., & Moran, C. J. (2014). The robustness of mine water accounting over a range of operating contexts and commodities. *Journal of Cleaner Production, 84*, 727–735. doi:10.1016/j.jclepro.2014.07.078

Daracon. (2010). Carrington barrier wall. Retrieved from http://www.daracon.com.au/What-We-Have-Achieved/Project-Profiles/Carrington-Barrier-Wall.aspx

David, K., & Dundon, P. (2010, October). *Mine water sharing – A way forward*. Paper presented at the Groundwater 2010, The Challenges of sustainable management, Canberra.

David, K., Timms, W., & Baker, A. (2014, October). *Detailed vertical profiling within the overburden strata, Sydney Basin, SE Australia*. Paper presented at the Geological Society of America, Vancouver.

Dragun, A. (1985). Water shortages in the Hunter Region of New South Wales. *Water International, 10*(4), 175–185. doi:10.1080/02508068508686349

Franks, D. M., Davis, R., Bebbington, A. J., Ali, S. H., Kemp, D., & Scurrah, M. (2014). Conflict translates environmental and social risk into business costs. *Proceedings of the National Academy of Sciences, 111*(21), 7576–7581. doi:10.1073/pnas.1405135111

Gleick, P., Christian-Smith, J., & Cooley, H. (2011). Water-use efficiency and productivity: Rethinking the basin approach. *Water International, 36*(7), 784–798. doi:10.1080/02508060.2011.631873

GRI. (2013). G4 sustainability reporting guidelines: Reporting principles and standard disclosures. Retrieved from https://www.globalreporting.org/resourcelibrary/GRIG4-Part1-Reporting-Principles-and-Standard-Disclosures.pdf

Gunningham, N., & Holley, C. (2010). *Bringing the 'R' word back: Regulation, environment protection and NRM*. Occasional Paper 3/2010. Canberra, Australia: The Academy of the Social Sciences in Australia.

Holley, C., & Sinclair, D. (2012). Compliance and enforcement of water licences in NSW: Limitations in law, policy and institutions. *The Australasian Journal of Natural Resources Law and Policy, 15* (2), 149–189.

Holley, C., & Kennedy, A. (2015, February). *Good governance and the energy water nexus*. Paper presented at the Frontiers of Environmental Law, Hobart.

Hussey, K., & Dovers, S. (2007). *Managing water for Australia*. Collingwood: CSIRO.

ICCM. (2012). Water management in mining – A selection of case studies. Retrieved from http://www.icmm.com/document/3660

Illman, W. (2014). Hydraulic tomography offers improved imaging of heterogeneity in fractured rocks. *Ground Water, 52*(5), 659–684. doi:10.1111/gwat.2014.52.issue-5

Irbaris. (2009). CDP water disclosure – The case for water disclosure. Retrieved from https://www.cdproject.net/en-US/Programmes/Documents/CDP_Water_Disclosure_PDF.pdf

Kennedy, A. (2016). *Environmental justice and land use conflict: The governance of mineral and gas resource development*. Abingdon: Earthscan.

Krogh, M., Dorani, F., Foulsham, E., McSorley, A., & Hoey, D. (2013). *Hunter catchment salinity assessment*. Sydney: Office of Environment and Heritage for the NSW Environment Protection Authority.

Kunz, N., & Moran, C. (2014). Sharing the benefits from water as a new approach to regional water targets for mining companies. *Journal of Cleaner Production, 84*, 469–474. doi:10.1016/j.jclepro.2014.02.053

Laurence, D. (2011). A guide to leading practice sustainable development in mining. Retrieved December 23, 2014, from http://www.industry.gov.au/resource/Documents/LPSDP/guideLPSD.pdf

Leong, S., Hazelton, J., Taplin, R., Timms, W., & Laurence, D. (2014). Mine site-level water reporting in the Macquarie and Lachlan catchments: A study of voluntary and mandatory disclosures and their value for community decision-making. *Journal of Cleaner Production, 84*, 94–106. doi:10.1016/j.jclepro.2014.01.021

MCA. (2012). Water accounting framework for the minerals industry user guide version 1.3. Retrieved from http://www.minerals.org.au/focus/sustainable_development/water_accounting

McGrath, C. (2014). One stop shop for environmental approvals a messy backward step for Australia. *Environmental and Planning Law Journal, 31*, 164–192.

Moolarben Coal. (2013). Annual environmental management report 2012-2013. Retrieved from http://www.moolarbencoal.com.au/__documents/major-project-approvals/environmental-managment-plans/annual-environmental-management-report/mco-aemr-2012-2013-main-doc.pdf

Morrison, J., & Schulte, P. (2010). Corporate water accounting. In P. Institute (Ed.), *United Nations Global Compact*. Oakland, CA: United Nations Environment Programme.

New South Wales Government. (2007). *State Environmental Planning Policy (Mining, Petroleum Production and Extractive Industries) 2007*. Retrieved April 14 2016 from http://www.legislation.nsw.gov.au/maintop/view/inforce/epi+65+2007+cd+0+N

NSW Department of Infrastructure Planning and Natural Resources. (2004). *A guide to the water-sharing plan for the hunter regulated river water source*. Sydney: Author.

NSW Department of Primary Industries. (2012). *NSW Aquifer Interference Policy: NSW Government policy for the licensing and assessment of aquifer interference activities*. Sydney: Author.

NSW Government. (2005). *Coal mining potential in the Upper Hunter Valley – Strategic assessment: NSW Department of Planning*. Sydney: Author.

NSW Government. (2010). Government prohibits open cut mining on Bickham site. Retrieved from https://www.google.com/webhp?sourceid=chrome-instant&rlz=1C1GGGE_enAU539AU539&ion=1&espv=2¡UTF-8#

NSW MC. (2014). Review of hunter river salinity trading scheme. Retrieved from http://www.nswmining.com.au/NSWMining/media/NSW-Mining/Publications/Submissions/140207_NSWMC-Submission_HRSTS-Review_FINAL.pdf

NSW Office of Water. (2014). Water accounting. Retrieved from http://www.water.nsw.gov.au/Water-management/Water-availability/Water-accounting/water-accounting

NWC. (2010). *Sustainable levels of extraction: National water commission position*. Canberra: Australian Government.

NWC. (2014a). *Australia's water blueprint national reform assessment 2014*. Canberra: Australian Government, National Water Commission.

NWC. (2014b). NWI and mining and unconventional gas. Retrieved from http://www.nwc.gov.au/publications/topic/water-planning/water-for-mining-and-unconventional-gas-under-the-national-water-initiative/nwi-and-mining-and-unconventional-gas

Owens, K. (2012). Strategic regional land use plans: Presenting the future for coal seam gas projects in New South Wales? *Environmental and Planning Law Journal, 29*(2), 113–128.

Parsons, R., Lacey, J., & Moffat, K. (2014). Maintaining legitimacy of a contested practice: How the minerals industry understands its 'social licence to operate'. *Resources Policy, 41*, 83–90. doi:10.1016/j.resourpol.2014.04.002

Prior, E. (2009). Water challenges for ASX100 companies – Adapting to increasing water-related risks. https://www.citigroupgeo.com/pdf/SAU06080.pdf.

Rau, G., Andersen, M., McCallum, A., Roshan, H., & Acworth, R. (2014). Heat as a tracer to quantify water flow in near-surface sediments. *Earth-Science Reviews, 129*, 40–58. doi:10.1016/j.earscirev.2013.10.015

Timms, W., Acworth, R., Hartland, A., & Laurence, D. (2012, September). *Leading practices for assessing the integrity of confining strata: Application to mining and coal seam gas extraction*. Paper presented at the International Mine Water Association, Bunbury, Western Australia.

Timms, W., Liu, H., & Laurence, D. (2013, November). *Design of a low permeability barrier (LPB) to limit seepage between a mine and a river*. Paper presented at the Water in Mining, Proceedings of the Australian Institute of Mining and Metallurgy, Brisbane.

Tonkin, C., & Timms, W. (2015). Styles of geological structures and fault-infill in the Southern Coalfields and implications for groundwater flow. *Journal of Research Projects Review, Mining Education Australia, 4*(1), 49–58.

Ulan Coal Mines Ltd. (2013). Annual Environmental Review. Retrieved from http://www.ulan coal.com.au/EN/ReportsandPublications/2013%20AER/2013_Annual%20Environmental% 20Review.pdf

Vink, S., Hoey, D., Robbins, S., & Roux, E. (2013, August). *Regulating mine water releases using water quality trading.* Paper presented at the Reliable Mine Water Technology, Proceedings of the International Mine Water Association Congress. Boulder, CO: Colorado School of Mines.

WEF. (2014). Global risks 2014 (9th ed.). Retrieved from http://www.weforum.org/reports/ global-risks-2014-report

Wheeler, S., Loch, A., Zuo, A., & Bjornlund, H. (2014). Reviewing the adoption and impact of water markets in the Murray–Darling Basin, Australia. *Journal of Hydrology*, *518*, 28–41. doi:10.1016/j.jhydrol.2013.09.019

Wood Mackenzie. (2013). Troubled waters ahead? Rising water risks on the global energy industry. Retrieved from http://public.woodmac.com/content/portal/energy/highlights/wk2_ Oct_13/Wood%20Mackenzie%20Troubled%20Waters%20Ahead_Abridged_Report.pdf

Non-discrimination and liability for transboundary acid mine drainage pollution of South Africa's rivers: could the UN Watercourses Convention open Pandora's mine?

Rémy Kinna

ABSTRACT

In 1997, South Africa became an inaugural party to the United Nations Watercourses Convention. With the convention entering into force in August 2014, South Africa is now bound by all of its provisions, including those relating to 'no significant harm' and liability for transboundary pollution. Article 32, the principle of 'non-discrimination', provides recourse for foreigners experiencing or under imminent threat of transboundary harm to seek compensation in the jurisdiction where the alleged harm originated. This article investigates the possibility under the convention of pursuing liability for transboundary acid mine drainage pollution originating in South Africa harming the Olifants-Limpopo Rivers.

Introduction

Acid mine drainage (AMD) is produced through the oxidization of pyrite when mine tailings and other mining by-products mix with water and are exposed to atmospheric air. The resulting AMD solution is not only highly acidic, but also mobilizes heavy metals, including copper and zinc, often in toxic concentrations (Ochieng, Seanego, & Nkwonta, 2010; Oelofse, 2008). The process of AMD is completed when this solution either: leaches back through the adjacent subsoil into surrounding aquifers; is pumped into surface waters to keep mine shafts from flooding and collapsing; or evaporates and falls as acid rain (Ochieng et al., 2010; Oelofse, 2008). AMD is thus extremely costly to prevent, contain and treat. Many of the chemicals and metals are 'bio-accumulators' that build up in flora and fauna to the detriment of both species and broader ecosystems (McCarthy & Humphries, 2013). Unsurprisingly, then, some scientists and water policy makers have characterized AMD as the "single most important environmental concern from mining activities" (Oelofse, 2008, p. 6; see also Ochieng et al., 2010; Republic of Mozambique, 2010).

AMD has been described as the "single greatest threat to South Africa's water-scarce environment" (Sharife, 2011). In turn, this has transboundary implications for South Africa, through its impact on some of the nation's most important international watercourses (Feris & Kotze, 2014; McCarthy, 2011).[1] Foremost among these is the

Olifants River tributary, which flows from South Africa and joins the Limpopo River mainstream in Mozambique (these are hereinafter referred to collectively as the Olifants-Limpopo Rivers). AMD is having severe local impacts upstream, as well as increasing transboundary impacts downstream, along the Olifants-Limpopo Rivers (Ashton, Love, Mahachi, & Dirks, 2001; Chilundo, Kelderman, & O'Keeffe, 2008; Republic of Mozambique, 2010). As a result, the cumulative impact of "AMD from several derelict and ownerless coal mines in the catchment creates tremendous long-term environmental liabilities" (Adler, Claassen, Godfrey, & Turton, 2007, quoted in Hobbs, Oelofse, & Rascher, 2008).

It is almost 20 years since South Africa became an inaugural signatory to, and ratified, the United Nations Convention on the Non-navigational Uses of International Watercourses (UNWC, 1997). The UNWC represents the key framework agreement governing international watercourses and was supported by over 103 nations when initially adopted in 1997 (Rieu-Clarke, Kinna, & Litke, 2013). Having eventually reached the necessary quorum of 35 ratifications with Vietnam ratifying the UNWC in May 2014, it entered into force on 17 August 2014. Its entry into force represented a critical advance in the global legal landscape regarding transboundary water law, in terms of both treaty law and customary international law. Hence, South Africa is now bound by the provisions of the UNWC, including the duty not to cause transboundary harm (Article 7) and all related obligations applicable to its international rivers.

Article 32 of the UNWC provides for recourse by persons, natural or juridical (for example, an incorporated company can be considered a 'juridical person' under the law), experiencing or under imminent threat of suffering transboundary harm related to an international watercourse to seek compensation within the jurisdiction where the alleged harm originates. As such, this provision allows access to judicial or other procedures, or a right to claim compensation or other relief with respect to such significant harm. Any application of Article 32 under the UNWC could have significant legal implications for transboundary water pollution from AMD originating from current, closed or abandoned mines of South Africa's international rivers. Furthermore, any potential invocation of Article 32 under the UNWC would be the first of its kind in the world given the UNWC's extremely recent entry into force.

Legal research has mainly focused on investigating state as well as private liability for AMD pollution in the context of domestic legislative regimes; this is especially the case in the context of South Africa, where the issue of who pays for the omnipresent legacy of mining pollution is especially thorny and complex (Feris, 2012; Feris & Kotze, 2014; Humby, 2013). However, there has been a dearth of research into the international legal nature of transboundary AMD pollution, particularly in the context of South Africa and shared watercourses *vis-à-vis* its neighbouring riparians.

With the UNWC entering into force, an opportunity arises to investigate whether the basic legal elements are present for its possible application regarding transboundary liability for South Africa's AMD pollution of its international watercourses. Using the case study of the Olifants-Limpopo Rivers, this article therefore seeks to provide a preliminary investigation into the potential for the UNWC to be invoked via Article 32 in seeking liability and even compensation for transboundary pollution from AMD originating in South Africa. Hopefully this sheds some light on the prospective role of the UNWC specifically, and international water law more generally, in opening

Pandora s box – or rather Pandora s mine in this context – on the feasibility of pursuing private legal remedies for AMD pollution across national borders within the domestic jurisdiction of a neighbouring riparian.

This article first explores the issue of AMD by highlighting the challenges faced by South Africa and pollution of freshwater resources, utilizing the case study of AMD causing transboundary impacts in Mozambique via the Olifants-Limpopo Rivers. It then outlines the UNWC, particularly its provisions related to transboundary harm and non-discrimination under Article 32, before highlighting the significance of its entry into force. Article 32 of the UNWC is subsequently applied in the context of South Africa's transboundary AMD pollution of an international watercourse via an investigation of whether the basic legal elements are present for Article 32 to be invoked by transboundary claimants in granting access to South Africa's domestic remedies. Finally, key questions for future research will be posed in view of the evolving legal landscape regarding liability for mining pollution of transboundary water resources in Southern Africa and globally.

AMD in South Africa: how mines are polluting the Olifants-Limpopo Rivers

South Africa's already scarce freshwater resources are under increasingly extreme stress. Regarding water availability alone, in October 2015, with the country facing its worst drought in over 30 years and 2.7 million households nationwide expected to face water shortages, the national government declared several provinces disaster areas; others are expected to follow (Essa, 2015). Water restrictions have already begun to be implemented in places such as Pretoria, yet the impacts go beyond simply household shortages given that the "hardest-hit provinces are also the country's most important food production areas" (Kings, 2015).

With such severe shortfalls now and in the foreseeable future, the scale of the freshwater crisis facing South Africa is further exacerbated by the rampant degradation of its dwindling supplies. AMD is one of the country's most highly publicized, politically sensitive, long-lasting and expensive forms of freshwater contamination, posing a threat to potable water but also industrial and agricultural sectors, many of which are water-intensive, including mining (Kings, 2015; Mathews, 2015). Cumulative environmental and health impacts, as well as monetary costs from AMD, have been allowed to mount over decades. Recent reports, though, suggest that South Africa's elected officials are finally taking heed of the far-reaching effects and overall gravity of the AMD situation (Kings, 2015; Mathews, 2015).

One of the country's transboundary rivers which is impacted by AMD pollution is the Limpopo River and its tributaries. Beginning in the north-eastern highlands of South Africa's Limpopo Province, the river stretches north along the border with Botswana and then arcs along the boundary with Zimbabwe before entering the south-eastern corner of Mozambique and continuing down to the country's capital, Maputo, where it reaches the Indian Ocean. Taking into account its 24 tributaries in total, the largest and most notable of which is the Olifants River (referred to as the Elefantes after it crosses the border into Mozambique at the Massingir Dam), the Limpopo River as a whole is one of the most important and largest transboundary rivers in the Southern

African Development Community (SADC) region (IMWI, 2008; Republic of Mozambique, 2010).

The river is also one of South Africa's most highly degraded. The significant environmental impacts witnessed both upstream and downstream are due to a range of factors affecting most of Southern Africa's major transboundary rivers, including "over-utilisation of water resources and pollution arising from high density urban settlements, *mining* and other industrial developments" (Malzbender & Earle, 2007, pp. 19–20, emphasis added). AMD mainly from closed, disused and abandoned mines which is occurring in the upper Olifants-Limpopo Rivers represents the most prominent case of this form of pollution in an international river originating within South Africa (Ashton et al., 2001; Republic of Mozambique, 2010).

Historical legacy of AMD in South Africa polluting the Olifants-Limpopo Rivers

The Olifants tributary itself has been described as "one of the hardest working, but also one of the most polluted, rivers in South Africa" (CSIR, 2011, p. 1). Indeed, 70% of the Olifants's total water usage in South Africa is allocated for agricultural irrigation (IMWI, 2008, p. 24). However, it is the historical water use and pollution issues associated with intensive coal and gold mining in and around Limpopo Province that have arguably had the most severe effects on its water quality and quantity, and which are now causing noticeable impacts downstream (Republic of Mozambique, 2010). Mining continues to be the largest economic contributor in this region of South Africa, where a legacy of poorly regulated opencast mining, conducted especially during the 1970s and 1980s, has left much of the environment adjacent to the Olifants River severely impacted, most prominently by AMD pollution (Mey & Van Niekerk, 2009). Lack of accountability continued even after the advent of democracy in 1994. Due to "inadequate enforcement and the lack of a regulatory system that provided for liability for environmental damage, the country faced deteriorating water quality and severe water pollution, including AMD" (Feris, 2012, p. 12). Mining pollution from AMD continued largely unchecked into the early 2000s due to critical implementation gaps in resourcing – both human and financial – as well as technical capacity for what were at the time some of the world's most advanced and ambitious regulations compared to the actual reality of the escalating crisis faced by not only Limpopo Province but mines throughout South Africa (Mey & Van Niekerk, 2009).

With many of these mines now abandoned or closed, and without proper maintenance or monitoring, uncontrolled discharges of water from mines are resulting in subsoil leaching into the local groundwater as well as unimpeded and/or untreated toxic acid mine spillage into surface waters, including along the Olifants-Limpopo Rivers (Hobbs et al., 2008; Oelofse, 2008). One significant example of this is the Western Basin gold mine, which is decanting into the Tweelopies River at the headwaters of the entire Limpopo River system and which also runs into another of its major tributaries, the Crocodile River (Turton, personal communication, 15 July 2015). This mine has been decanting since 2002 at approximately 20–60 million litres per day with mining fluid contaminated with heavy metals from AMD, and until recently samples taken have also been rich in radioactive uranium (Turton, personal communication, 15 July 2015). Whilst this single example is truly alarming, the most worrying reality of today's crisis is that the threat of AMD for water resources in South Africa has been known for decades" (Feris, 2012, p. 4).

AMD impacts on freshwater ecosystems

In terms of harmful impacts, AMD severely diminishes water quality and concurrently causes broader environmental impacts such as soil contamination with toxic metals, modifying the biochemistry of marine species and aquatic habitats (Earthlife, 2010; Saving Water South Africa, 2010). Significantly, based on existing knowledge and practices, it is estimated that such impacts in the most severe circumstances could persist for decades if not hundreds of years to come, thus threatening the sustainability of affected surface and aquifer freshwater resources and their associated environments for the foreseeable future (Ochieng et al., 2010; Oelofse, 2008). In the context of the gold and coal mines along the Olifants-Limpopo Rivers, this is disturbing to contemplate. "Ultimately, pyrite in the rocks in these mining areas will be fully oxidized and AMD will cease. There is no indication as to how long this will take, but the problem is likely to persist for centuries rather than decades" (McCarthy, 2011, p. 5).

AMD originating in South Africa causing transboundary impacts

The downstream impacts of AMD pollution occurring in South Africa have increasingly been recognized by various scientific studies (Chilundo, Kelderman, & O'Keeffe, 2008; Mey & Van Niekerk, 2009; Roux, Oelofse, & de Lange, 2010). This culminated in a South African Department of Water Affairs report expressing "serious concern" (DWA, 2010, p. iv) regarding AMD pollution of the Olifants River and other domestic water resources. Arguably, though, it took a highly publicized incident of pollution of the Olifants River, which occurred in 2008, with the sudden deaths of crocodiles in the Kruger National Park section of the tributary in South Africa, bordering the Massingir Dam in Mozambique, to provoke a serious response from the South African government (Republic of Mozambique, 2010). The department conducted an Inter-Ministerial Report in 2010 investigating the effects of AMD in and around the most industrialized areas of South Africa. The study, whilst focusing on mining areas proximate to Johannesburg, also mentioned that "severe water related problems, including numerous AMD decants have been reported in the Mpumalanga Coal Fields" (DWA, 2010, p. 33).

Around the same time, a group of over 30 government scientists were conducting ongoing research into the impacts of AMD on the Olifants River since 2009. Their initial risk assessment report, released in 2011, found that the upper catchment of the Olifants River in South Africa is highly contaminated with "heavy and trace metal ions and sulphate, attributable to abandoned mining and industrial activities" (CSIR, 2011, p. 1). Significantly, this study identifies and isolates an "acute need ... to counter the current situation of poor water quality and to halt or (preferably reverse) the existing pattern of progressively increasing eutrophication and contamination ... to prevent the propagation of adverse water quality impacts further down the Olifants River" (p. 1).

The Mpumalanga Coal Fields, in the upstream area of the Olifants River, are a significant region for non-point source AMD pollution of the Olifants River impacting downstream areas. Downstream and across the border in Mozambique, it has also been reported by scientists that dams such as Massingir are acting as pollution sinks (Roux, Oelofse, & de Lange, 2010), trapping heavy metals and toxic chemicals in their stagnant water and silt which accumulate to increasingly dangerous concentrations over time (Mey & Van Niekerk, 2009). Therefore, the critical issue of the AMD pollution of the Olifants River tributary of the Limpopo River that is occurring in South Africa provides an ideal scenario to examine this intersection of transboundary harm and the potential for establishing breaches of the duty of no significant harm as well as transboundary liability for harm caused by AMD.

South Africa's UNWC obligations regarding transboundary pollution liability

South Africa signed the UNWC on 13 August 1997 and ratified it on 26 October 1998, thus making the nation a party to the convention. Entry into force means that South Africa is now bound by all of its provisions. Its main purpose is to "ensure the utilisation, development, conservation, management and protection of international watercourses and the promotion of the optimal and sustainable utilisation thereof for present and future generations ... taking into account the special situation and needs of developing countries" (Preamble). In this respect, the UNWC codifies the three main principles of customary international law on watercourses: equitable and reasonable utilization (Articles 5 and 6); prevention of significant harm (Article 7); and prior notification for planned measures that are likely to cause transboundary impacts, such as hydropower dams (Article 12) (Bearden, 2010).

The UNWC's cornerstone is the principle of reasonable and equitable use. States must "utilise an international watercourse in an equitable and reasonable manner" (Article 5), and this enjoys legal primacy over *all* other principles in the UNWC (Salman, 2015). Its subordination of the obligation not to cause significant harm (discussed below) was a source of debate during the lengthy negotiations and many draft articles that ultimately led to the UNWC, but is now generally agreed (Rieu-Clarke & Gooch, 2009–2010). Such extensive negotiations on key points of law such as this form part of the reason why the UNWC is widely regarded as "the most comprehensive and important codification of international watercourse law" (Bearden, 2010, p. 805) and contains many principles which are widely recognized as customary law (McCaffrey, 2001). As such, it has long been commonly viewed as the most authoritative source of law on transboundary watercourses, which is now further bolstered by its entry into force as an enforceable treaty (Litke & Rieu-Clarke, 2015).

Several UNWC provisions specifically relate to the issue of transboundary harm from pollution; most notably, Article 7 codifies the duty to do "no significant harm" with respect to international watercourses. The most disputed of all the principles in the UNWC, as noted above, it obliges states "to take all appropriate measures" (Article 7) to utilize an international watercourse so as not to cause significant harm to another riparian state (McCaffrey, 2001). Other obligations directly related to the duty not to cause significant harm and inter-connectedly the principle of equitable and reasonable utilization include Part IV of the UNWC, comprising Articles 20–26, which is explicitly concerned with the protection, preservation and management of watercourse ecosystems and environments; Articles 27 and 28, which detail duties whereby states must immediately notify neighbouring riparians of harmful conditions and emergency situations, including those caused by human conduct such as industrial accidents and/or pollution incidents that could potentially cause transboundary impacts within their territories; and finally, Article 32, which provides the right for legal persons to pursue private recourse, legal or otherwise, including compensation for the imminent threat or occurrence of transboundary harm, in the watercourse state where the harm has allegedly originated. Taken together, the above package of interrelated and complementary provisions oblige riparian *states* to cooperate in taking all reasonable measures and act within their jurisdictions to ensure they do not cause harm to an international watercourse and its related environments within another riparian state's borders. The focus of this analysis, however, now turns specifically to the application of Article 32 regarding *private* remedies – those that are not between sovereign states – for the imminent threat or actual occurrence of transboundary harm and its operation in conjunction with Article 7.

Obligation not to cause significant harm

Article 7(1) sets out the requirement that states take "all appropriate measures" to prevent causing "significant harm". The obligation to take all appropriate measures is one of due diligence, whereby a state must implement whatever measures are reasonable to avoid transboundary harm occurring. In terms of what constitutes *significant* harm to a watercourse state, Article 1 of the 2000 SADC Revised Protocol on Shared Watercourses, which was modelled on the UNWC, states that significant harm is defined as "non-trivial harm capable of being established by objective evidence without necessarily rising to the level of being substantial". Significant harm can thus be taken to mean a level of harm that is "higher than merely perceptible or trivial (which would be considered insignificant), but can be less than severe or substantial" (Rieu-Clarke, Moynihan, & Magsig, 2012, p. 120). Furthermore, the harm must be "more than just an 'adverse effect' – it engenders a real impairment of a use, with a detrimental impact of some consequence upon the environment or the socioeconomic development of the harmed state (e.g. public health, industry, property, agriculture)" (p. 120). Hence, what constitutes significant harm must be evaluated and decided based on *objective* evidence and decided on a case-by-case basis; the burden of proof rests with the state allegedly causing the harm to establish that such harm is still equitable and reasonable (Rieu-Clarke et al., 2012; Wouters, Vinogradov, Allan, Jones, & Rieu-Clarke, 2005).

Article 7(2) stipulates that where significant harm is caused, states will have "due regard for the provisions of Articles 5 and 6" regarding the principle of equitable and reasonable utilization, including the factors to be taken into consideration in determining what is equitable and reasonable. The duty in Article 7(2) also includes consulting with an affected state in order "to eliminate or mitigate such harm and, where appropriate, to discuss the question of compensation". This added obligation of discussing the question of compensation is rarely examined in regard to the duty to do no harm under the UNWC and will not be dealt with here in detail as it concerns state-to-state claims. However, the obligation of no significant harm and potential compensation for a breach thereof do coalesce in relation to private remedies and transboundary harm under Article 32 of the UNWC.

Non-discrimination and the right of foreigners to pursue liability for transboundary harm

Article 32 deals with transnational *private* liability, specifically, the right of individual citizens and private legal entities to pursue liability and compensation for transboundary damage against other legal persons, including the state, in the jurisdiction where the alleged harm originated. This principle of equality of access for all private citizens/entities to transboundary legal remedies for harm suffered, or the threat of harm, which originates outside their sovereign territory, is referred to as non-discrimination. Non-discrimination therefore seeks to allow any foreign legal persons alleging harm equal access to judicial pathways in the sovereign jurisdiction where the harm is alleged to have originated.

The origins of the principle of non-discrimination can be traced back to the landmark decisions of the 1938 and 1941 *Trail Smelter Arbitrations (US v Canada)*. Collectively lasting over a decade from 1926, these matters were actually initiated on the grounds that American farmers alleged that they had suffered transboundary harm from a Canadian smelter and were not granted the right to access Canada's domestic legal system in order to pursue their claims (Percival, 2010). These milestone arbitrations led to legal developments in terms of access to justice for transboundary harm whereby non-discrimination has since been adopted as a fundamental international legal principle in Article 2(6) of the 1991 United Nations Economic Commission for Europe (UNECE) Convention on Environmental Impact Assessment in a Transboundary Context ('Espoo Convention'); the Preamble and Article 9 of the 1992 UNECE Convention on the Transboundary Effects of Industrial Accidents; Articles 1, 3(9) and 9 of the 1998 UNECE Convention on Access to Information, Public Participation in Decision-Making and Access to Justice in Environmental Matters ('Aarhus Convention'); Chapter 3, Article 3 of the 1992 Las Leñas Protocol on Jurisdictional Cooperation and Assistance; and Article 15 of the 2001 International Law Commission (ILC) Draft Articles on Prevention of Transboundary Harm from Hazardous Activities (Boyle, 2012). Collectively, these legal agreements evince that the principle is now widely recognized and incorporated into environment-related treaties.

Non-discrimination is also codified in international environmental liability instruments. Notable examples include Article 19 of the 1993 Council of Europe Convention on Civil Liability for Damage Resulting from Activities Dangerous to the Environment, and specifically for transboundary waters, Article 13 of the 2003 UNECE Protocol on Civil Liability and Compensation for Damage Caused by the Transboundary Effects of Industrial Accidents on Transboundary Waters. However, many of these liability agreements are not yet in force due to states' unwillingness to be bound to international liability standards (Percival, 2010).

The International Court of Justice (ICJ, 2010), in the *Pulp Mills on the River Uruguay* case, decided in 2010, found that Argentinian and Uruguayan citizens both enjoyed equal access to the other nation's judicial systems and accompanying legal remedies and procedures with respect to transboundary harm (Birnie, Boyle, & Redgwell, 2009; Rieu-Clarke et al., 2012). Combined with a growing corpus of private international law matters which recognize the "jurisdiction of national courts to hear cases involving transboundary harm to extraterritorial plaintiffs" (Boyle, 2012, p. 638), it is therefore evident that now more than ever, victims of transboundary pollution already have rights in international law which they can exercise within the legal system of the polluting state" (p. 635).

In the UNWC, non-discrimination is codified in Article 32, which provides that:

> Unless the watercourse states concerned have agreed otherwise for the protection of the interests of persons, natural or juridical, who have *suffered or are under a serious threat of suffering significant transboundary harm* as a result of activities related to an international watercourse, a watercourse state shall not discriminate on the basis of nationality or residence or place where the injury occurred, in granting to such persons in accordance with its legal system, access to judicial or other procedures, or *a right to claim compensation* or other relief in respect of significant harm caused by such activities carried on in its territory [emphasis added].

Article 32 specifically grants "access to judicial or other procedures, or *a right to claim compensation*" (emphasis added). This leads to the matter of establishing a right to claim liability and/or compensation for transboundary harm or serious threat thereof, in accordance with the legal system where the harm originated, which will be explored below. Another crucial legal element of Article 32 is that *no actual* transboundary harm is necessary. Rather, transboundary claimants from a foreign watercourse state are required to establish only that they are "under a *serious threat* of suffering transboundary harm", meaning that it is likely to be caused by activities originating in the state where they seek to access redress via domestic courts. Regarding this definition, commentary in the 1994 International Law Commission Draft Articles on International Watercourses states that persons "under a serious threat of suffering transboundary harm" (ILC, 1994, p. 132) applies where "the harm is prospective in nature" (p. 133). Crucially for foreign claimants seeking liability for transboundary harm, this significantly lowers the legal threshold to establishing the mere impending threat of harm, albeit serious in nature, rather than actual harm occurring (although that still remains additional/concurrent grounds for enforcing Article 32).

Why investigate non-discrimination and liability in the context of transboundary AMD harm?

Article 32 of the UNWC represents a potentially landmark legal development for both the government and private sector in South Africa in terms of potential exposure to liability and compensation, as well as for foreign legal persons, natural or juridical, who have been harmed, or are under imminent threat of harm, by AMD pollution originating from South Africa. As Rieu-Clarke, Moynihan and Magsig (2012, p. 229) point out:

> Once the UN Convention has entered into force, its incorporation into the national legal system of a state party should imply an automatic adjustment of the national legal system (or the creation of additional implementing legislation by states as per their domestic legislation) to implement the international legal obligations under Article 32 and *the procedural right to access could be directly invoked by private foreign nationals without delay* [emphasis added].

Applying the principle of non-discrimination contained in Article 32 of the UNWC therefore lends itself as a topic for legal analysis on several fronts, particularly in the context of South Africa.

First, the provision for non-discrimination as contained in Article 32 is unique amongst South Africa's international agreements pertaining to its transboundary watercourses. No equivalent provision is codified in the overarching SADC Revised Protocol of 2000, which governs all international watercourses in the region and which South Africa is a party to, along with all the other members of the SADC but Zimbabwe (Kinna, 2015). This is particularly significant because the SADC Revised Protocol is an amended version of the original SADC Protocol on Shared Watercourses of 1995, which was explicitly revised with the intention of aligning it with the UNWC – so much so that the Revised Protocol repeats much of the UNWC verbatim. Moreover, none of South Africa's multilateral agreements over its major international watercourses such as the Limpopo, Orange-Senqu and Incomati-Maputo river basins contain any provision similar to the principle of non-discrimination codified in the UNWC's Article 32 (Kinna, 2015).

Second, legal recourse and liability for harm are normally predicated, not only within domestic jurisdictions for nationals of that country but especially in the case of persons who are seeking to claim international civil liability, on establishing *causal injury*: a causal link between the wrongful conduct and the resulting actual harm (Percival, 2010). Whilst the causal connection element must still be satisfied, under the UNWC principle of non-discrimination a mere *imminent threat* of harm can be satisfactory, rather than proving *actual* injury occurred. Therefore, the potential for invoking Article 32 based on the imminent threat of harm to those seeking liability and even compensation for transboundary harm cannot be overlooked. It also raises the possibility of foreign claimants' not only seeking liability and compensation for current/historic actual harm, but also utilizing judicial or other mechanisms to guarantee that the serious threat of *future* harm is avoided in the state where the harm to a transboundary watercourse is originating. Based on South Africa's past, present and, likely, future contamination of water resources within its territory by AMD, especially given the bioaccumulative nature of its environmental impacts downstream, Article 32 poses significant legal liability questions based on its respective temporal applications (Feris & Kotze, 2014).

Finally, as international environmental treaties progressively adopt non-discrimination as a central principle, "transboundary claimants can be empowered to act as part of the enforcement structure of international environmental law by giving them access to the same information, decision-making processes, and legal procedures as nationals" (Boyle, 2012, p. 635). A parallel trend is that "private parties have been aggressively pursuing transnational environmental tort litigation" (Percival, 2010, p. 49). Therefore, any such utilization of the principle of non-discrimination under Article 32 of the UNWC would contribute to a growing trend over the last 30 years of legal cases initiated by, or on behalf of, individuals and communities directly affected by transboundary environmental harm (Boyle, 2012).

Can UNWC Article 32 be invoked to seek liability for transboundary AMD harm?

In light of the entry into force of the UNWC, South Africa is now legally obliged to adhere to all its provisions. Using the case study of transboundary AMD harm to the

Olifants-Limpopo Rivers in Mozambique, the section below investigates whether the basic legal elements exist to satisfy the Article 32 and related provisions in order to access South Africa's domestic laws regarding liability from mining pollution, specifically AMD.

Does the UNWC's entry into force provide recourse for transboundary AMD claims?

The following legal analysis applies the basic facts of the case study regarding transboundary harm from AMD pollution of the Olifants-Limpopo Rivers system to the principle of non-discrimination in Article 32 of the UNWC and related provisions, plus international case law.

Has a person, natural or juridical, 'suffered or are [they] under a serious threat of suffering significant transboundary harm'?

In order to establish grounds for a claim under Article 32, there are must have been a threat or actual suffering of transboundary harm. Concerning the particular impacts of AMD on the Olifants-Limpopo Rivers and effects threatened and/or evinced in Mozambique, changes in water quality can constitute a basis for establishing harm to the natural environment where "harmful contaminants traverse a border through a transboundary watercourse or aquifer, thereby impacting the environment, habitats, species, or dependent ecosystems of another riparian State" (UNEP, 2011, p. 47). In the *Pulp Mills* case, the International Court of Justice was asked to consider the differences in levels of certain pollutants recorded downstream in Argentina before and after the construction of the Orion (Botnia) mill in Uruguay. The court sought to determine whether the impact of the mill's pollutant discharge on the water quality of the Uruguay River constituted transboundary harm for the purposes of establishing a breach of the obligation "to prevent pollution and preserve the aquatic environment" (ICJ, 2010, pp. 229–265). Despite acknowledging differences in certain concentrations, after evaluating the available scientific evidence and expert testimony related to each distinct chemical, the court deemed the differences between levels recorded before and after operation of the mill too insignificant either to constitute harm or to attribute to the mill operations.

In the instance of AMD within South Africa causing transboundary harm, research conducted in Mozambique has reported a significant increase over time in the pollutants derived from upstream activities such as mining, whereby heavy metals such as zinc, iron, copper and cadmium were present in the water at all sampled sites along the lower Olifants River and Massingir Dam at levels higher than Mozambican national water quality standards, and some, including zinc, copper and cadmium, above World Health Organization standards (Chilundo et al., 2008). It has been determined that these concentrations of heavy metals in the lower Limpopo River are derived from "sediment transport along the river coming from upstream mining areas" (Chilundo et al., 2008, p. 659), specifically upstream mining activities in South Africa.

Below the confluence where the Olifants flows into the Limpopo, there is a further deterioration in water quality, which is estimated to be the "residual effect of the mining activities in the upper reaches of the catchment" (Republic of Mozambique,

2010, p. 17). Furthermore, water quality in the Limpopo River Basin was found deteriorated and not meeting the guidelines for potability" (Chilundo et al., 2008, p. 664). As a result, the major impacts of AMD on the Limpopo River have been identified as increased levels of heavy metal toxicants (particularly in the Olifants sub-catchment) and ions (specifically in the Changane sub-catchment) in the water (p. 664).

Despite limited scientific evidence of actual harm along the Olifants River in Mozambique directly linkable to point-source water contamination from mining operations in South Africa, existing research shows that the effects of AMD have been slowly propagating downstream. It is now at the point where immediate action is required in order to prevent and abate upstream transboundary harm from AMD pollution sources in South Africa causing further long-term impacts in Mozambique (McCarthy, 2011; Mey & Van Niekerk, 2009). Indeed, AMD is a recognized threat that could more broadly affect South Africa's other riparian neighbours along the larger Limpopo River basin, Zimbabwe and Botswana (Plaut, 2013). However, due to "the particular local conditions, the problems in the Olifants ... are huge by comparison and pose a serious threat to future generations of South Africans" (McCarthy, 2011, p. 6).

In sum, given that (1) preventing AMD from coal mines can be so complex that the possibility of preventing uncontrolled AMD decanting from rehabilitated opencast mines may not exist, and that (2) there is "still a large amount of unmined coal in the Olifants catchment and many prospecting and mining applications await approval ... which will undoubtedly lead to further increases in the pollution loads in the future" (McCarthy, 2011, p. 5), it is clear that the already degraded water quality of the Olifants River will continue to deteriorate from AMD due to coal mines in South Africa (p. 6). Consequently, the severe and often irreversible impacts of AMD originating in South Africa will continue to propagate downstream, including in Mozambique along the Olifants-Limpopo Rivers into the foreseeable future. Therefore, taking into consideration the acute impacts on water quality described above which have already been recorded downstream across the border in Mozambique, especially at the Massingir Dam as noted earlier, it is reasonable to proceed on the basis that this element (threat of and/or actual significant transboundary harm) may be sufficiently established.

Level of transboundary harm caused by AMD must be "significant"

"Significant harm" is not defined in the text of the UNWC; one must take note of the *travaux preparatoires* (preparatory materials involved in the drafting of the official legal instrument) and supporting documentation for the text of the UNWC in seeking to establish a definition. Under Article 7 of the UNWC, the term "significant" is not equated with "substantial" (Rieu-Clarke et al., 2012). Rather, as regards harm, "While such an effect must be capable of being established by objective evidence and not be trivial in nature, it need not rise to the level of being substantial" (Rieu-Clarke et al., 2012, p. 64). Hence, significant harm under the UNWC can reasonably be interpreted as being in line with the SADC Revised Protocol, insofar as it is "*non-trivial harm* capable of being established by objective evidence without necessarily rising to the level of being substantial" (Article 1, emphasis added). Not only are these toxic heavy metals extremely harmful to the environment, even in small concentrations, but they are also often bio-accumulators, which infiltrate living tissue and can also persist for centuries (Ochieng et al., 2010; Oelofse, 2008; Republic of Mozambique, 2010). As a result, this element appears to be satisfied based on the discussion above and earlier in relation to the scientific evidence of impacts on water quality from heavy metals recorded in Mozambique.

Is the harm "a result of activities related to an international watercourse"?

Under Article 32, there are a number of identifiable barriers to establishing that the threat or occurrence of harm is indeed "a result of activities related to an international watercourse", warranting an application of the specific facts of this case study in seeking to prove causation. Isolating a point source of pollution in South Africa, such as a mine or group of mines, producing the AMD pollution that has caused, or is threatening to cause, transboundary impacts within Mozambique would be at worst implausible, or at best, extremely difficult. Transboundary harm via the Olifants-Limpopo Rivers could feasibly be coming from any, or every, mine producing AMD within this basin catchment. Therefore, establishing causation would probably require claimants to demarcate those sources of AMD which they allege are causing and/or threatening harm to the international watercourse based on a more precise set of criteria, such as the geographic area with the highest concentration of AMD-producing mines, or those polluting mines with the same current or past owners. As will be touched upon later in relation to domestic laws, pursuing a claim against the South African government for a failure to protect international watercourses under its jurisdiction might be a 'catch-all' way of circumventing the need for claimants to delineate such specific criteria.

Adding complexity, the Olifants-Limpopo Rivers are awash with pollutants from multiple sources, including chemicals from agricultural fertilizer runoff (CSIR, 2011; IMWI, 2008). Hence, establishing that past, previous or future mining activities producing AMD *solely* caused the particular combination of chemical contaminants resulting in harm within Mozambique will be challenging, to say the least. Nevertheless, this barrier may foreseeably be overcome thanks to the mounting research outlined above relating to harm within Mozambique having been identified as increased levels of heavy metal toxicants which are scientifically attributed specifically to upstream mining activities in South Africa, particularly AMD pollution (Ashton et al., 2001; Chilundo et al., 2008; Mey & Van Niekerk, 2009; McCarthy, 2011; Republic of Mozambique, 2010).

Where the thorny issue of proving causation arises, there is no discernible answer. Establishing causation is simply explained in Percival's (2010, pp. 42–43) analysis regarding the distinct challenges of liability for harm with respect to point versus non-point pollution sources:

> In situations where large, single sources of pollutants, such as smelters, caused visible environmental damage, the common law tort of nuisance could provide some measure of redress to plaintiffs. But in a modern world awash in pollutants from multiple sources, the difficulty of proving causal injury has made common law liability too crude a vehicle to compensate those exposed to environmental hazards. To be sure, when a particular toxic substance, such as asbestos, causes 'signature injuries' uniquely tied to exposure to it, the

causation conundrum can be overcome. Yet even in the case of asbestos, because exposure to this deadly substance caused fatal diseases with a long latency period, liability was imposed only decades after exposure to the products containing it.

Although this presents a basic understanding of the difficulties in establishing causal harm, it does pinpoint a key barrier to transboundary liability, whereby even in scenarios such as that of asbestos – which involved direct contact, with a specific point source of pollution, resulting in signature injuries – satisfactorily establishing causation is not easily assured and/or may take a long time for those injuries to propagate in order to meet the burden of proof. In the *Pulp Mills* case, causation with respect to transboundary harm to an international watercourse was relatively straightforward to establish as the alleged transboundary harm was from point-source chemicals produced by the Orion (Botnia) pulp mill located directly on the Uruguay River (ICJ, 2010). Drawing a parallel to this case, unless mine(s) in South Africa are directly spilling AMD-polluted water into the Olifants-Limpopo Rivers, it could be difficult to show causation. Ultimately, establishing that those mines previously were, currently are, or might in the future be producing AMD in South Africa are definitively *activities related* to harm of the Olifants-Limpopo Rivers will be at the crux of any transnational claims seeking to invoke Article 32. Given that the water quality impacts attributed to AMD which have been recorded in Mozambique noted above were taken from locations along the Olifants-Limpopo Rivers (including the Massingir Dam where the Olifants crosses the border) this analysis can therefore proceed on the basis that causation here might reasonably be satisfied.

Can foreigners establish a private claim for liability under South Africa's domestic laws?

Ultimately, establishing such a claim for liability and compensation will rest upon national laws recognizing transboundary pollution of international watercourses as an offence within the domestic jurisdiction where the legal matter is being pursued. This is where international water agreements cross into the domestic legal systems of riparian states. While the basic legal elements of Article 32 may potentially be satisfied based on the preliminary analysis above, the UNWC merely provides an international legal mechanism for foreign legal persons to seek to access the legal (and also administrative if available) regimes within the domestic jurisdiction of the neighbouring riparian where the harm is alleged to have originated. In this regard, "the non-discrimination principle requires the polluting state to treat extra-territorial nuisances no differently from domestic nuisances" (Boyle, 2012, pp. 639–640). Therefore, any potential claim for liability and compensation under the UNWC will always inherently be predicated on there being valid grounds for those legal persons, natural or juridical, to be granted standing within the national legal system to pursue such an action.

Feris and Kotze (2014) provide a comprehensive analysis of all of South Africa's domestic laws governing liability for AMD pollution. The first point of reference and the foundation of the national legal framework is the 1996 Constitution of the Republic of South Africa. Under Section 24(a), the Constitution guarantees that everyone has the right to an environment that is not harmful to their health and well-being. It also incorporates the principle of inter-generational equity in regard to everyone's having the right to environmental protection for present and future generations. Section 24(b) subsequently instructs the state to adopt reasonable legislative and other measures to prevent pollution, promote conservation, and secure ecologically sustainable development and use of natural resources. In doing so, such mechanisms should also seek to promote justifiable economic and social development. Lastly, and specifically regarding water, Section 27(1)(b) stipulates that all South Africans have the right to "sufficient food and water", whereby "the State must take reasonable legislative and other measures, within its available resources, to achieve the progressive realisation of each of these rights" (Section 27(1)(c)).

The practical legal effect with regard to the impacts of AMD pollution within South Africa is that Section 24(a) grants rights to all citizens to have their environment (taken to mean the immediate environment with which they interact in the course of their daily lives, as well as that under the territorial sovereignty of the state as a whole) and personal health protected from harm caused by AMD (Feris, 2012). Section 24(b) imposes a distinct duty upon the state to take reasonable measures to prevent, reduce and control any AMD pollution as well to ensure that natural resources, particularly freshwater in this case, are conserved in current and future mining developments. Considered together with Section 27(1)(b), "these rights require of the state to ensure that water is conserved and protected and that sufficient access to the resource is provided" (Feris, 2012, p. 12).

To give statutory effect to the constitutional objectives outlined above, several legislative regimes have been enacted, including:

> the NEMA [National Environmental Management Act 107 of 1998], which regulates the protection of all environmental resources, including water; the WSA [Water Services Act 108 of 1997], which regulates access to potable water supply services; the NWA [National Water Act 36 of 1998], which ensures the management, protection and conservation of water resources; and the National Environmental Management: Waste Act (NEM:WA); which provides for the management of waste. South African law also provides for the regulation of mining in the MPRDA [Mineral and Petroleum Resources Development Act 28 of 2002] (Feris & Kotze, 2014, pp. 2120–2121).

Together, this statutory framework, along with subsequent amendments such as the National Environmental Management Amendment Act 14 of 2009, encompasses specific provisions for preventing, minimizing and remediating water pollution from past, existing and future mining activities, including liability for harm, which *inter alia* applies to AMD (Feris & Kotze, 2014). For lack of space, this article cannot detail all the relevant statutes applicable to liability in the context of transboundary harm from AMD. Nevertheless, from Feris and Kotze's examination of all relevant provisions, it is clear that a comprehensive set of laws exist which can impose statutory-administrative as well as criminal liability (on company directors) on legal persons – including companies and government agencies – found responsible for AMD pollution. Crucially, any of these statutes could potentially be applicable to foreign nationals using Article 32 of the UNWC to initiate a liability claim via South Africa's courts. Investigating in detail all of the possible domestic liability avenues available to foreign claimants provides scope for further research.

Role of citizens and NGOs in future claims and advancing global liability standards

In terms of legal standing to pursue AMD liability under South Africa's domestic law, under the principle of non-discrimination, "the argument that transboundary victims come within the jurisdiction or control of the polluting state can be made, is consistent with existing human rights law, and is supported by developments in international environmental law" (Boyle, 2012, p. 642), such as the Aarhus Convention. Notwithstanding, and as noted earlier, in this new era of global environmental law, NGOs are fighting liability battles on behalf of local communities with multi-national companies (Percival, 2010, p. 49). Moreover, "research has shown that [NGOs] tend to have high success rates in enforcement actions and public interest litigation" (Boyle, 2012, p. 626). Hence, a number of environmental and human rights NGOs in South Africa who focus on protecting the rights of communities from mining contamination, including AMD, may also hold the key to pursuing domestic legal claims on behalf of foreign nationals.

The Centre for Environmental Rights (CER) is one such NGO, which acts on behalf of civil society and community organizations regarding mining pollution. In September 2015, CER instituted legal proceedings for a coalition of groups against the Mineral Resources Minister in Pretoria's High Court following his granting of a coal mining right to an Indian-owned mining company. This right allows Atha-Africa Ventures to mine inside a Protected Environment which is "the source of three major rivers – the Tugela, the Vaal, and the Pongola – that support downstream water users who will be affected if the source of those rivers is compromised" (Kotze, 2015). The legal basis of the claim is that the "Minister's decision to grant the right was unlawful because coal mining in such a strategically important area would result in unacceptable pollution, ecological degradation and damage to the environment". One of the coalition representatives commented that "the mining right was granted without having regard to downstream water users dependent on the water". Significantly, an environmental impact assessment conducted by the mining company found that "seepage of acid mine drainage from the mine into the freshwater system is highly likely".

While CER's action is based on the legal rights of organizations and communities within South Africa, it certainly mirrors key elements of the potential liability claims examined earlier stemming from AMD transboundary pollution along the Olifants-Limpopo Rivers causing impacts in Mozambique. It also comes soon after the recent high-profile decision in *Harmony Gold Mining Company Ltd v Regional Director: Free State Department of Water Affairs and Others*. In that case, decided in June 2012, a mining company was held liable, based on an unfulfilled government directive, for the costs of pumping and treating AMD pollution from its former mine, despite the mine and the land upon which it was situated having already been sold several times over to different companies, some of which were now bankrupt (Feris & Kotze, 2014; Humby, 2013). This verdict was subsequently upheld in February 2014 when the Constitutional Court (South Africa's highest court) refused any leave to appeal (Odendaal, 2014). With South Africa's drought pushing limited available freshwater resources to the breaking point, NGOs such as CER will in all likelihood be increasingly involved in using the law to protect communities from existing or potential impacts of mining pollution, especially AMD (Earthlife, 2010; Odendaal, 2014). Whether any future lawsuits are initiated for downstream communities within *and* outside South Africa, only time will tell.

Whether it is foreign citizens or local NGOs who potentially pursue liability for transboundary AMD pollution in South Africa, either would contribute to a growing global trend, whereby "private actors and nongovernmental organizations are driving the development of new legal and non-legal strategies to protect the environment" (Percival, 2010, p. 37). Thus, as more and more victims of transboundary environmental harm pursue liability and compensation for injury via private transnational litigation, we are arguably seeing a global pattern emerge where environmental liability norms are being developed 'from the bottom up', mostly by NGOs and communities. In doing so, these environmental and community advocates are "blurring lines that traditionally separated conceptions of domestic and international law and public and private law" (p. 37). Hence, any invocation of UNWC Article 32 by claimants in the context of transboundary water resources and liability for mining pollution could represent further advancement in environmental liability standards, not only in South Africa but globally. Doing so may open 'Pandora's mine' on future similar claims regarding mining pollution of water resources that cross national borders around the world. Moreover, it might alter the perception of the previously overlooked legal significance of Article 32 of the UNWC, as well as a greater role for, and broader application of, the principle of non-discrimination in international water law.

Transboundary rivers, AMD and South Africa's water: dawn of a new liability era?

It is evident from the central argument of this article, along with the rapidly expanding body of research on this subject, that South Africa has reached a crucial juncture regarding AMD pollution impacts on its increasingly scarce and degraded freshwater resources. This urgent issue encompasses transboundary harm to one of its major international watercourses, the Limpopo River and its tributaries, particularly its largest and most degraded, the Olifants River. Undoubtedly, the impacts of AMD on the water quality and the aquatic environment of the Olifants-Limpopo Rivers are approaching, or may have already reached, a threshold whereby there will be long-term and quite possibly irreversible impacts within the upper reaches of this international watercourse within South Africa. If this trend continues, the weight of scientific evidence clearly suggests that certain heavy metal toxicants from AMD will continue to propagate downstream into Mozambique and lead to extremely harmful levels similar to those already being recorded in South Africa.

Significant legal developments have occurred recently for South Africa transboundary water resources, most notably the entry into force of the UNWC in August 2014. This has opened the possibility for foreign claimants to use the principle of non-discrimination to pursue liability and compensation for transboundary harm due to AMD pollution originating in South Africa. Whether such claims are actually pursued, only time will tell. However, the issue of liability and compensation for AMD pollution of water resources, especially the legal complexity and practical difficulties surrounding liability for historic AMD pollution, is one that is not going away anytime soon – not unlike the pervasive impacts of AMD contaminating the Olifants-Limpopo Rivers.

Note

1 A 'watercourse' (used interchangeably with 'river' in this article) is defined as "a system of surface and ground waters constituting by virtue of their physical relationship a unitary whole and normally flowing into a common terminus" (UNWC, 1997, Article 2); 'international watercourses' have parts which are situated in different states.

Acknowledgements

The author wishes to extend his sincere thanks to Prof. Loretta Feris (University of Cape Town) for her helpful comments on earlier drafts of this article, as well as the feedback of the peer reviewers. All errors remain the author's own. Finally, the author would like to acknowledge the constant support and encouragement of Cristina Morgante.

References

Adler, R., Claassen, M., Godfrey, L., & Turton, A. R. (2007). Water, mining, and waste: An historical and economic perspective on conflict management in South Africa. *The Economics of Peace and Security Journal, 2*(2), 32–41. doi:10.15355/epsj.2.2.33

Ashton, P. J., Love, D., Mahachi, H., & Dirks, P. H. G. M. (2001). An overview of the impact of mining and mineral processing operations on water resources and water quality in the Zambezi, Limpopo and Olifants Catchments in Southern Africa. *Contract Report to the Mining, Minerals and Sustainable Development* (Southern Africa). *Project, by CSIREnvironmentek, Pretoria, South Africa and Geology Department, University of Zimbabwe, Harare, Zimbabwe. Report No. ENV-P-C 2001-042.* Retrieved from: http://pubs.iied.org/pdfs/G00599.pdf

Bearden, B. L. (2010). The legal regime of the Mekong River: A look back and some proposals for the way ahead. *Water Policy, 12*, 798–821. doi:10.2166/wp.2009.060

Birnie, P. W., Boyle, A. E., & Redgwell, C. (2009). *International law and the environment* (3rd ed.). Oxford, UK: Oxford University Press.

Boyle, A. E. (2012). Human rights and the environment: Where next? *The European Journal of International Law, 23*(3), 613–642. doi:10.1093/ejil/chs054

Chilundo, M., Kelderman, P., & O'Keeffe, J. H. (2008). Design of a water quality monitoring network for the Limpopo River Basin in Mozambique. *Physics and Chemistry of the Earth, Parts A/B/C, 33*(8–13), 655–665. doi:10.1016/j.pce.2008.06.055

Council of Europe Convention on Civil Liability for Damage Resulting from Activities Dangerous to the Environment, 1993, adopted on 21 June 1993 (not yet in force).

CSIR. (2011). Risk assessment in pollution in surface waters in the upper Olifants River System: Implications for aquatic ecosystem health and the health of human users of water. *Interim report to the Olifants River Forum – Executive Summary.* Retrieved from: http://www.orf.co.za/PDF/Risk%20Assessment%20of%20Pollution%20in%20Surface%20Waters_March%202011.pdf

Department of Water Affairs (DWA) South Africa. (2010). Mine Water Management in the Witwatersrand Gold Fields with Special Emphasis on Acid Mine Drainage. *Report to the Inter-Ministerial Committee on Acid Mine Drainage.* Retrieved from: http://www.dwaf.gov.za/Documents/ACIDReport.pdf

Earthlife. (2010, January 19). Latest acid mine drainage crisis calls for a constructive response from civil society. *Earthlife Africa.* Retrieved from: http://www.chroniclesa.co.za/index.php?

view=article&catid=1:latestnews&id=712:environmental-disaster-flowing-from-the-west-rand &format=pdf

Essa, A. (2015, November 4). South Africa in midst of 'epic drought'. *Al Jazeera*. Retrieved from: http://www.aljazeera.com/news/2015/11/south-africa-midst-epic-drought-151104070934236. html

Feris, L. (2012). The public trust doctrine and liability for historic water pollution in South Africa. *Law, Environment and Development Journal*, 8(1), 1–20.

Feris, L., & Kotze, L. J. (2014). The regulation of acid mine drainage in South Africa: Law and governance perspectives. *Potchefstroom Electronic Law Journal*, 17(5), 2105–2163.

Harmony Gold Mining Company Ltd v Regional Director: Free State Department of Water Affairs and others (*Harmony Gold II*) 68161/2008 (29 June 2012).

Hobbs, P., Oelofse, S. H. H., & Rascher, J. (2008). Management of environmental impacts from coal mining in the upper Olifants River catchment as a function of age and scale. *International Journal of Water Resources Development*, 24(3), 417–431. doi:10.1080/07900620802127366

Humby, T.-L. (2013). Commentary – The spectre of perpetuity liability for treating acid water on South Africa's goldfields: Decision in *Harmony II. Journal of Energy & Natural Resources Law*, 31(4), 453–466. doi:10.1080/02646811.2013.11435343

Integrated Water Management Institute (IMWI). (2008). Baseline Report: Olifants River Basin in South Africa. *Waternet Website*. Retrieved from: http://www.waternetonline.ihe.nl/challenge program/AR15%20CP17-Baseline%20study%20Olifants.pdf

International Court of Justice (ICJ). (2010). *Pulp Mills on the River Uruguay* (*Argentina v Uruguay*) (*Pulp Mills Case*) ICJ Advisory Opinion (Order of 20 April, 2010) ICJ Reports. The Hague: International Court of Justice.

International Law Commission (ILC). (1994). Draft Articles on the Law of the Non-Navigational Uses Of International Watercourses and Commentaries Thereto and Resolution on Transboundary Confined Groundwater. Retrieved from: http://untreaty.un.org/ilc/texts/instru ments/english/commentaries/8_3_1994.pdf

International Law Commission (ILC). (2001). Draft Articles on Prevention of Transboundary Harm from Hazardous Activities. Retrieved from: http://legal.un.org/ilc/texts/instruments/ english/commentaries/9_7_2001.pdf

Kings, S. (2015, August 28). Departments admit to acid water slip-up. *Mail & Guardian*. Retrieved from: http://mg.co.za/article/2015-08-27-departments-admit-to-acid-water-slip-up

Kinna, R. (2015, May). *From pioneer to pragmatist or pariah: Are South Africa's recent water law developments compatible with its obligations under the UN Watercourses Convention?* Paper presented at the International Water Resources Association World Water Congress, Edinburgh, Scotland.

Kotze, C. (2015, September 18). Centre for Environmental Rights institutes legal action on coal mining right. *Mining Review*. Retrieved from: http://www.miningreview.com/centre-for-environmental-rights-institutes-legal-action-on-granting-of-coal-mining-right/

Litke, A., & Rieu-Clarke, A. (2015, February 2). The UN Watercourses Convention: A milestone in the history of international water law. *Global Water Forum*. Retrieved from: http://www. globalwaterforum.org/2015/02/02/the-un-watercourses-convention-a-milestone-in-the-his tory-of-international-water-law/

Malzbender, D., & Earle, A. (2007). The impact and implications of the adoption of the 1997 UN Watercourse Convention for countries in Southern Africa. *WWF International -Global Freshwater Programme*. Retrieved from: http://www.acwr.co.za/pdf_files/WWF_RA_UN% 20Convention.pdf

Mathews, C. (2015, October 26). Water Affairs works to clear acid mine drainage. *Business Day Live*. Retrieved from: http://www.bdlive.co.za/national/2015/10/26/water-affairs-works-to-clear-acid-mine-drainage

McCaffrey, S. (2001). The Contribution of the UN Convention on the law of the non-navigational uses of international watercourses. *International Journal of Global Environmental Issues*, 1(3/4), 250–263. doi:10.1504/IJGENVI.2001.000980

McCarthy, T. S. (2011). The impact of acid mine drainage in South Africa. *South African Journal of Science, 107*(5/6), 1–7. doi:10.4102/sajs.v107i5/6.712

McCarthy, T. S., & Humphries, M. S. (2013). Contamination of the water supply to the town of Carolina, Mpumalanga, January 2012. *South African Journal of Science, 109*(9/10), 1–10. doi:10.1590/sajs.2013/20120112

Mey, W. S., & Van Niekerk, A. M. (2009, October). *Evolution of mine water management in highveld coalfields.* Paper presented at the International Mine Water Conference, Pretoria, South Africa. Retrieved from: http://www.imwa.info/docs/imwa_2009/IMWA2009_Mey.pdf

Ochieng, G. M., Seanego, E. S., & Nkwonta, O. I. (2010). Impacts of mining on water resources in South Africa: A review. *Scientific Research and Essays, 5*(22), 3351–3357.

Odendaal, N. (2014, February 19). ConCourt rules on acid mine drainage liability. *Mining Weekly.* Retrieved from: http://www.miningweekly.com/article/concourt-rules-on-acid-mine-drainage-liability-2014-02-19

Oelofse, S. (2008, March). Mine water pollution - Acid Mine Decant, Effluent and Treatment: A Consideration of Key Emerging Issues that May Impact the State of the Environment. *Department of Environmental Affairs & Tourism.* Retrieved from: http://www.anthonyturton.com/admin/my_documents/my_files/Mine_Water_Pollution.pdf

Percival, R.V. (2010). Liability for environmental harm and emerging global environmental law. *Maryland Journal of International Law, 25,* 37–63.

Plaut, M. (2013, May 23). Johannesburg: Who pays for a century of mining? Retrieved from: http://martinplaut.wordpress.com/2013/05/23/johannesburg-who-pays-for-a-century-of-mining

Protocol of Las Leñas on Jurisdictional Cooperation and Assistance in Civil, Commercial, Labor, and Administrative Matters, 1992, adopted on 27 June 1992. Retrieved from: www.mercosur.org.uy/espanyol/snor/normativa/decisiones/DEC0597.htm

Republic of Mozambique. (2010). Joint Limpopo River Basin study – Scoping phase – Final report. *Report on behalf of Limpopo Basin Permanent Technical Committee.* Retrieved from: www.icp-confluence-sadc.org/project/docs/publicfile?id=190

Republic of South Africa, *Constitution of the Republic of South Africa* (Constitution) (1996).

Republic of South Africa, *Mineral and Petroleum Resources Development Act* (MPRDA) (28 of 2002).

Republic of South Africa, *National Environmental Management Act* (NEMA) (107 of 1998).

Republic of South Africa, *National Environmental Management Amendment Act* (NEMAA) (14 of 2009).

Republic of South Africa, *National Environmental Management: Waste Act* (*NEM:WA*) (59 of 2008).

Republic of South Africa, *National Water Act (NWA)* (36 of 1998).

Republic of South Africa, *Water Services Act* (108 of 1997).

Rieu-Clarke, A., & Gooch, G. (2009–2010). Governing the tributaries of the Mekong-The contribution of international law and institutions to enhancing equitable cooperation over the Sesan. *Pacific McGeorge Global Business & Development Law Journal, 22,* 193–224.

Rieu-Clarke, A., Kinna, R., & Litke, A. (2013). UN Watercourses Convention: Online User's Guide, Centre for Water Law, Policy and Science, University of Dundee. Retrieved from: www.unwatercoursesconvention.org/

Rieu-Clarke, A., Moynihan, R., & Magsig, B. O. (2012). UN watercourses convention user's guide. *IHP-HELP Centre for Water Law, Policy and Science (under the auspices of UNESCO),* Dundee, UK.

Roux, S., Oelofse, S., & De Lange, W. (2010, September). *Can SA afford to continue polluting its water resources? – With special reference to water pollution in two important catchment areas.* Paper presented at the CSIR 3rd Biennial Conference: Science Real and Relevant, Pretoria, South Africa. Retrieved from: http://researchspace.csir.co.za/dspace/bitstream/10204/4262/1/Roux_2010.pdf

Salman, S. M. A. (2015). Entry into force of the UN Watercourses Convention: Why should it matter? *International Journal of Water Resources Development, 31*, 4–16. doi:10.1080/07900627.2014.952072

Saving Water South Africa. (2010, June 16). Acid mine drainage threat could persist for several hundred years. Retrieved from: http://www.savingwater.co.za/2010/06/16/10/amd-threat-persist-for-hundred-years/

Sharife, K. (2011, May 2). South Africa: Companies profit from toxic water. *The Africa Report.* Retrieved from: http://www.minesandcommunities.org/article.php?a=10876

Southern African Development Community Revised Protocol on Shared Watercourses (SADC Revised Protocol), 2000, opened for signature 7 August 2000 (entered into force 22 September 2003).

Trail Smelter Arbitrations (US v Canada). UN Reports of International Arbitral Awards (16 April 1938 and 11 March 1941) Volume III, 1905–1982. Retrieved from: http://untreaty.un.org/cod/riaa/cases/vol_III/1905-1982.pdf

United Nations Convention on the Non-Navigational Uses of International Watercourses (UNWC), 1997, opened for signature 21 May 1997, entered into force 22 August 2014 (36 ILM 700).

United Nations Economic Commission for Europe Convention (UNECE) on Access to Information, Public Participation in Decision-making and Access to Justice in Environmental Matters ('Aarhus Convention'), 1998, adopted on 25th June 1998, entered into force 30 October 2001.

UNECE Convention on Environmental Impact Assessment in a Transboundary Context ('Espoo Convention'), 1991, adopted 25 February 1991, entered into force 10 September 1997.

UNECE Convention on the Transboundary Effects of Industrial Accidents, 1992, adopted on 17 March 1992, entered into force 19 April 2000.

UNECE Protocol on Civil Liability and Compensation for Damage Caused by the Transboundary Effects of Industrial Accidents on Transboundary Waters, 2003, opened for signature 21 May 2003 (not yet in force).

United Nations Environment Programme (UNEP). (2011). *The greening of water law: Managing freshwater resources for people and the environment.* Kenya: UNON Publishing Services.

Wouters, P. K., Vinogradov, S., Allan, A., Jones, P., & Rieu-Clarke, A. (2005). Sharing transboundary waters: An integrated assessment of equitable entitlement: The legal assessment model. *International Hydrological Programme* (IHP) *of the United Nations Educational, Scientific and Cultural Organization (UNESCO),* Dundee, Scotland. Retrieved from: http://www.chinainternationalwaterlaw.org/pdf/resources/LegalAssessmentModel-.pdf

A pilot study of the Social Water Assessment Protocol in a mining region of Ghana

Anastasia N. Danoucaras, Alidu Babatu Adam, Kathryn Sturman, Nina K. Collins and Alan Woodley

ABSTRACT

The Social Water Assessment Protocol (SWAP) is a tool consisting of a series of questions on 14 themes designed to capture the social context of water around a mine site. A pilot study of the SWAP, conducted in Prestea-Huni Valley, Ghana, showed that some communities were concerned about whether the ground-water was potable. The mining company's concern was that there was a cycle of dependency amongst communities that received treated water from the mining company. The pilot identified potential data sources and stakeholder groups for each theme, and gaps in themes, and suggested refinements to questions to improve the SWAP.

Introduction

Mining companies have become aware of the need to have better governance of social impacts due to mine water use. In both developed and developing countries, there is better regulation and soft laws surrounding the mitigation of the environmental and social impacts of mining, as well as the requirement for community development (Dupuy, 2014; McNamara, 2013; Sethi, Lowry, Veral, Shapiro, & Emelianova, 2011). Water is increasingly understood as not just an environmental issue but also a social one. The human right to water and sanitation was explicitly recognized by the United Nations General Assembly in 2010 (UNGA, 2010). Against this backdrop, mining companies have an obligation to ensure that their activities do not infringe upon this right (Kemp, Bond, Franks, & Cote, 2010).

A key concern related to the potential impacts of mining is the depletion of a community's essential freshwater resources. Excessive withdrawals and/or pollution of freshwater by mine operations have the potential to reduce the volume and quality of water for other uses. Mining companies can also block communities' access to water through diversions, which is where a naturally occurring stream or creek is rerouted to allow access to valuable ore or to provide land for mining-related infrastructure.

Beyond immediate consumption requirements, people need sufficient water for both livelihood and domestic uses (Goff & Crow, 2014), and water resources also fulfil cultural purposes (Perreault, 2014). The management of water, when it is viewed only as a resource to sustain human life and ecosystems, focuses on technical specifications such as volume and quality, without taking account of other social, cultural and economic values (Perreault, 2014). Mine water management must be thought of in broader terms than merely considering the abstractions and discharges due to a mine site. For this reason, it is important that mining companies understand the social water context in which they are operating.

Collins and Woodley (2013) developed the Social Water Assessment Protocol (SWAP), a series of approximately 60 questions on 14 themes relating to the community's interactions with water sources. The general purpose of the SWAP is to help mine site staff understand the social water context of their operations and to complement existing water reports. The questions are designed to span climate conditions, social groupings, economic drivers and political situations. The questions relate to water and the infrastructure for the region, legislative issues, human rights, health, gender roles, and the various types of water uses: domestic, amenity, livelihood, spiritual, cultural and recreational. There are numerous questions for each topic within each theme, and it is not expected that all questions would be required to be asked in every application of the tool. Whilst there are questions to understand the physical water withdrawals from the environment, the interaction between water and ecosystem function is outside the scope of the tool.

Collins and Woodley (2013) proposed that the SWAP outputs can provide a contextual statement for the Minerals Council of Australia's Water Accounting Framework (MCA, 2012), thereby supporting the use of the framework in discussing mine water management with communities (Kemp et al., 2010). Water account reporting is done regularly at intervals determined by the mine site. Linking the SWAP to the Contextual Statement of the Water Accounting Framework means the SWAP will also be undertaken regularly, which will ensure that companies will capture changes in the social water context over time (Collins & Woodley, 2013).

Social impact assessment is the primary process through which mining companies understand and address the social impacts of a project (MMSD [Mining, Minerals and Sustainable Development], 2002). Whilst social impact assessment should, in theory, be an ongoing process, in practice, it requires other supporting processing and tools to ensure that social impacts are continually monitored and managed to deliver long-term positive outcomes *throughout* the life cycle of the mine site (Franks & Vanclay, 2013; MMSD, 2002). The SWAP is one such supporting tool. Another company-specific tool is Anglo American's Socio-Economic Assessment Toolbox, which explores a number of water-related impacts, including water supply, water pollution, water for economic production, links between water and health, access to water, and quality of water (Anglo American, 2007). The Global Environment Management Initiative's Collecting the Drops tool is another. This tool is a questionnaire containing a total of 18 questions that considers how the business engages with stakeholders regarding water (GEMI, no date). While both these tools are valuable in their own right, the SWAP focuses specifically on the social dimensions of water at the site level and from the point of view of the surrounding communities, and arguably allows a greater depth and breadth of investigation than other tools.

The pilot study obtained answers to the SWAP questions from a desktop review and a field trip to the mine site region of the case study. Whilst the general aim of the

SWAP is to understand the uses and values of water for stakeholders of the region, the primary aim of this study was to pilot the SWAP and improve its usability by refining the questions where appropriate. This article outlines findings from the SWAP answers, suggests amendments to the questions and identifies data sources and stakeholder groups to facilitate data collection for future applications of the SWAP.

Case study area

The pilot was conducted in a region surrounding a mine site in the Prestea-Huni Valley District of Ghana. Prestea has been mined for gold since 1873 via private and state-owned interests. In the 1980s, the Ghanaian government encouraged privatization and foreign investment in the mining sector (Hilson & Yakovleva, 2007). JCI Ltd operated the Prestea mine from the 1990s, but closed it in 1998 due to unprofitability. It was then taken over by a coalition of the Prestea workforce called Prestea Gold Resources. In March 2002, Golden Star Bogoso Prestea Limited partnered with Prestea Gold Resources to manage the mine, but mining at Prestea has not happened since then (Golden Star Resources, 2013). When the underground mine was operational, there were over 4000 people employed by the company, and when it closed, many turned to artisanal and small-scale mining (ASM), known in Ghana as *galamsey* (CHRAJ, 2008). The Bogoso gold mine does not have the same legacy issues. Mining has been occurring at Bogoso since 1935. Golden Star bought the mine in 1999, and surface mining is still active (Golden Star Resources, 2013).

Method

Table 1 gives a description of the thematic areas and the data sources that were used for this study. The term "context" when used in reference to the SWAP means the study area, which must be established for each application of the SWAP to a mining region. Applying the SWAP means to obtain answers to the applicable questions across the 14 themes; therefore, the context does not change from theme to theme. The criteria for delineating the context for the SWAP analysis are determined through initial scoping of the site's water footprint and through engagement with mine company personnel. The boundary of the context should not be constrained by the mine site concession; water-related impacts and water catchments should be considered when deciding the boundary of the SWAP context. This includes whether the community itself perceives water-related impacts due to the presence of the mine. The area of the mining concession was the context for this particular application of the SWAP because there were many communities in close proximity to the active and inactive mines. The concession is described in the first theme, Physical Water Snapshot. The purpose of this theme is to understand the geography of the region and the water footprint of the mining concession, but the context can be broader than the concession. It is anticipated that in other applications of the SWAP, the boundary of the context would not usually align with the mining concession.

To obtain the data to answer the SWAP questions, the authors undertook a desktop study and a field trip to the mine site region. The desktop review was of readily available information from the Internet, water-related legislation and policies, and reports obtained

Table 1. Topics within each theme of the Social Water Assessment Protocol, and data sources used to obtain the results.

Theme	Sample of topics within theme	Data sources for current study
Theme 1: Physical water snapshot	Area of the mining concession (not necessarily the same as the boundary of the context), water resources of the region, water quality of the region, and the interaction of the mine site with the water resources	http://www.gsr.com/, extracts from Golden Star Environmental Management Plan, social responsibility and environmental affairs representative from corporate office of Golden Star, community interviewees
Theme 2: Climate	Climate conditions of the region such as rainfall and evaporation, what effect it has on water resources and the mine site, and the likelihood of drought and flooding	Golden Star Environmental Management Plan, report provided by district representative of Ministry of Food and Agriculture
Theme 3: Water supply and infrastructure	Water infrastructure, responsibility of supply, sufficiency of supply and access to supply	Social responsibility and environmental affairs representative from corporate office of Golden Star, community interviewees
Theme 4: Local amenities	Water used for local amenities such as schools, hospitals, transport and hydroelectricity	Community interviewees, representatives from health at community and district levels
Theme 5: Domestic uses	Water used for domestic purposes – water supply, affordability, volumes, quality, and water resources shared with the mine site	Community interviewees, district planning officer
Theme 6: Formal and informal economy	The formal and informal industries that consume water, the industries that interact with water, the size of the industries, the quantity and quality of the water that is used by the industries	Report provided by district representative of Ministry of Food and Agriculture, community and mine site interviewees, field observations
Theme 7: Indigenous peoples	If there are Indigenous people in the context, how they interact with water, the cultural significance of the water sources, and water resource sharing between the mine site and indigenous people	Not applicable
Theme 8: Social, cultural and spiritual	Identifies the particular water sources that have social (i.e. recreational), spiritual, religious or cultural significance, how people interact with the water sources on these bases and the values that the community holds for the water sources	Community interviewees including chiefs and elders
Theme 9: Human rights	Extent to which the government in the context acknowledges that water is a human right, human rights abuses, identification of marginalized groups with respect to sufficient water access	Desktop review (see references) and interview with a representative from the Commission for Human Rights and Administrative Justice in Tarkwa
Theme 10: Gender	Differences between men and women in responsibility for water collection, end uses of water, customary practices that involve water, the water sources they draw on, water management at the household/community/government levels, water accessibility and sanitation access	Community interviewees and district planning officer
Theme 11: Health	Access to clean water, access to sanitation, water-related diseases, health risks associated with artisanal mining	Representatives from health at community and district levels, community interviewees
Theme 12: Other key stakeholders	Other stakeholders within the context that have not been previously identified, water volumes used, access to water, interaction between stakeholders	Community interviewees
Theme 13: Interaction between stakeholders	Cooperation between stakeholders, water trading, water-related conflicts, site's engagement with the other local stakeholders	Mining company representatives and community interviewees
Theme 14: Legislation, policy, politics	Water policy of the context, contribution of stakeholders to water policies, enforcement of policy and legislation, water resource authorities, water restrictions, water entitlements and licencing	Desktop research of legislation, regulations and policy

from key informants, such as Golden Star s Environmental Management Plan. The field trip deployed key informant interviews and focus group discussions. Primary data from interviews and focus group discussions may reflect perceptions or opinions. The team conducting the interviews consisted of a social scientist, a water scientist, and a member of the local community, who acted as an interpreter.

Prior to the field trip, contact with Golden Star representatives delineated the context and provided initial answers to SWAP questions. During the field trip, conducted between 20 October and 30 October 2013, other stakeholders were interviewed and further questioning of mine site representatives was conducted. In total, four mine site representatives were interviewed from community relations, the environmental department and the corporate office. Participants were drawn from communities within the mine site concession. Bogoso and Prestea are the two major towns within the Golden Star Bogoso mine site concession. Currently, the area surrounding Bogoso has active mining, so interviews were concentrated around Bogoso and its nearby smaller communities: Dumasi, Bepo, Chujah, Joaben, Kojokrom, Eshereso and Komsono. Interviews were also conducted at Prestea, Bondaye (near Prestea) and the smaller community of Himan (between Bogoso and Prestea), although large-scale mining was no longer conducted near these sites. There are other smaller communities within the mining concession, but it was impractical to visit all of them.

Seventy-five participants were interviewed over 25 interviews; 18 interviews were with one to two individuals (four company representatives, seven key informants and eight local authorities), and there were seven focus group discussions (one women's group, two men's groups, one community-based organization, one farmers' group, one group of chiefs and sub-chiefs and one group of artisanal small-scale miners). Nearly a third of the respondents were female. The researchers targeted the questions of the SWAP themes appropriate to the particular group or individual being questioned; thus, not all questions were asked of each participant. For instance, elders and chiefs were asked about the theme regarding cultural significance of the water bodies. The breadth of topics covered by the tool meant that some questions were not relevant in that context, and therefore they were not asked of the interviewees. The interviews were conducted in a semi-structured or conversational manner.

Results

The full set of answers from the SWAP can be found as part of a larger document on the International Mining for Development website (Danoucaras, Babatu, & Sturman, 2014); the results have been condensed for this manuscript to focus on the pilot study.

Theme 1: Physical water snapshot

The concession occupies an area of 1600 ha and encompasses 16 towns and villages: Adamanso, Bepo, Kumsono, Appiatse, Eshereso, Bogoso, Dumasi, Kojokrom-Akokobedi Aburo, Juaben, Himan, Chujah, Kwamenuampa, Dumaa, Prestea, Himan and Bondaye, under six divisional chiefs. A map of the concession area can be found in the full report (Danoucaras, Babatu, & Sturman, 2014). Bogoso (the district capital, population 36,000)[1] and Prestea (population 31,000)[2] are towns; the rest are small rural communities. The native language of the people is Akan (Twi and Ahanta dialects).

A representative from the corporate office of Golden Star provided relevant parts from the environmental management plan to answer the questions within this theme. The company extracts groundwater for mining purposes and does not use any surface water other than rainfall and runoff. The groundwater is high in dissolved salts associated with the mineralogy of the area. The representative of the mining company believes that the groundwater resource is sufficient for the quantity being taken.

Within the Bogoso concession there are two main river catchments: the Mansi River and the Bogyiri River. Both discharge into the Ankobra River. Within the Prestea concession area, there are four tributaries that also enter the Ankobra River. The Ankobra and Mansi Rivers are the prominent surface water resources within the concession.

Multiple interviewees reported that the streams and creeks in the concession are severely degraded and polluted through ASM and farming and are rarely used for any domestic purpose. The main pollutants include sediments and mercury from ASM, coliforms from weak sanitation systems, and pesticides from farming. For most of the local communities that used to rely on surface water, Golden Star has provided access to groundwater with the installation of wells. The company confirmed that regulated diversions of creeks impacted access to surface water in communities such as Dumasi and Chujah, so Golden Star provides potable water for these communities.

Theme 2: Climate conditions

The region is within a rainforest zone, with a bimodal distribution of annual rainfall. There is a major wet season occurring annually from April to June, a minor dry season in July to August, a smaller wet season in September to November, and a main dry season in December to February. The mean maximum temperature was 32 °C, and the mean minimum temperature 22 °C, over the six-year period from 2005 to 2011. Relative humidity is typically in the range of 70–90% throughout the year.

Theme 3: Water supply and infrastructure

The source of the domestic supply is groundwater. In Ghana, the Community Water and Sanitation Agency (CWSA) and various local district assemblies are responsible for planning and delivering potable water infrastructure in small rural communities, whilst the Ghana Water Company Ltd is responsible for urban water supply. The Ghana Water Company supplies water to small sections of the Bogoso and Prestea communities.

District officials who were interviewed explained that there are plans to fix and/or expand water facilities in every community, but these efforts are challenged by budget constraints and difficulty accessing some communities with borehole drillers. The efforts of the CWSA to provide safe water to local communities are complemented by private operators. In the Bogoso area, a private water supplier called Christ the King Water Company operates a mechanized borehole system and provides water to about two-third of Bogoso.

In the last 20 years, Golden Star has provided about 100 tailor-made water facilities (boreholes/wells with pumps) for the communities, depending on population size and ability of the drilling teams to get to the proposed site of the borehole. These facilities have been provided both for impact mitigation from previous stream diversions and for social development purposes. The company is worried that community reluctance to pay user fees for water has created a dependency syndrome. Golden Star wants to hand over responsibility of these water facilities to the community to ensure long-term sustainability.

Smaller communities such as Komsono, Esheroso, Kojokrom and Appiatse access potable water through boreholes/wells fitted with pumps at communal points. The Dumasi and Chujah communities, whose surface water resources were diverted by mine operations, rely exclusively on Golden Star. The mining company has placed communal water tanks around these communities, and treated water from the site is hauled regularly.

Some respondents from the communities of Bepo, Appiatse, Komsono and Himan had concerns with taste and colour of the groundwater. These communities instead rely on sachet water (potable water in small plastic bags) for drinking as they are concerned about the presence of a layer of film that is evident when the water is allowed to stand. The community members referred to this film as "oily". The researchers were informed that the groundwater is naturally high in iron, and desktop research after the field trip suggested that the residue may be due to iron-oxidizing bacteria (National Ground Water Association, 2009).

In general, community interviewees reported that they obtain sufficient water to meet basic needs; water for bathing and cooking is drawn from communal water sources, and drinking water is obtained from potable water suppliers.

Theme 4: Local amenities

There are several social amenities in the surveyed communities, including schools, clinics, hospitals, banks and churches. The majority of the facilities rely on communal water facilities for water. A few health facilities, including the Bogoso and Prestea government hospitals, have running water from mechanized borehole systems.

Sanitation facilities for the disposal of both solid and liquid waste are poor. With the exception of two public water-closet toilets in the Bogoso township that are connected to running water from boreholes, all other public toilets are vault-chamber or KVIP (Kumasi ventilated-improved pit) types and are in a poor condition. Disposal of solid waste in most communities is open-field disposal and has the risk of contaminating streams through runoff.

Theme 5: Domestic uses

The field data revealed that most people draw on groundwater from boreholes and wells for domestic uses, including cooking, washing and cleaning. Where the groundwater appears to have the oily residue previously mentioned, the majority of interviewees indicated that they drink sachet water. Apart from a significant proportion of houses (estimated at 40% of households) in Bogoso that are connected to running water, households in all other surveyed communities collect their water from communal standpipes, boreholes, wells, and less commonly streams. Community members also collect and store rainwater for supplementary use during the rainy season.

Access to water in the area is generally good and relatively accessible, and facility-user fees are economically affordable. In general, most people travel a distance of less than 400 m, with a turnaround time of 15 minutes or less, to collect water from facilities. The time spent collecting water is longer during the dry season (30 minutes to an hour). The cost of water varies according to facility and community size. Whilst community members reported that the current water user fees are affordable, willingness to pay varied amongst community members, depending on the provider. In communities where water facilities are provided by the local District Assembly, Ghana Water Company Ltd or private operators, community members are willing to pay for water. In contrast, communities that are provided water by Golden Star are reluctant to pay for water.

In all the surveyed communities, there is no known social or legal restriction of any group of people in terms of access to water.

Theme 6: Formal and informal economy

The structure of the local economy is a mix of agriculture, mining, service and micro-industrial activities. According to the 2010 Ghana National Population and Housing census, 44.1% of the economically active population in the district are employed in agriculture and related activities; 18.2% in mining and quarrying; and 13.6% in whole-sale and retailing (Ghana Statistical Service, 2014). Other sectors include service and small-scale manufacturing industries. The mineralogy of the district is endowed with gold deposits, which makes it attractive to both large-scale and artisanal gold mining operators. In addition to employment provided by regulated large-scale mining companies like Golden Star Resources, the majority of youth are engaged in ASM (*galamsey*). Interview data suggest that there are about six ASM sites in the context. Cumulatively, each site employs approximately 200 people, made up of groups operating in smaller units. Livestock rearing is very rare, and where it exists, consists of only a few birds and small ruminants such as sheep and goats. Currently, there are about 36 fish farmers in the area, but pollution of surface water means that the fish farming can be safely done only with dug-out fish ponds.

With regard to sector interactions with water, the majority of the interviewees consider both formal mining and ASM the major water users in the region. Agriculture can be a significant consumer of water, but it is rainfed. The land does not require irrigation; therefore agriculture does not extract surface or groundwater. Formal mining extracts groundwater for both direct mine operations (for example ore processing, cooling systems) and auxiliary services such as catering and camp management services. ASM activities create pollution, with major impacts on surface water, such as the Mansi and Ankobra Rivers as well as nearby streams.

Theme 7: Indigenous peoples

This theme is not relevant in this context as there are no people that are recognized as indigenous people in Ghana. This was corroborated by local and company sources.

Theme 8: Social, cultural and spiritual

Whilst the majority of surface water resources in the area are unsafe for domestic use, some rivers and streams remain culturally significant and are regarded as symbols of community history, clan identity, social cohesion and spiritual protection.

Interviewees narrated the spiritual connection they share with the gods that are associated with nearby streams and rivers. For example, the Bogo stream at Bogoso, Nsuo Kofi stream at Prestea, Mansi River and Eshere stream at Eshereso, and Achesua and Afiada Nsuo streams at Dumasi are regarded as gods. These gods have cultural protocols which community chiefs/priests strive to preserve and protect. At Eshereso, oral tradition prescribes that no person is allowed to collect water from the Mansi River on a Tuesday or the Eshere on a Wednesday. Men can swim or bathe, but women age 18 and over are not allowed to visit the Mansi River on those days. Those who violate these protocols suffer punishment as prescribed by the chief. Similarly, the chief and people of Dumasi perform annual libation and sacrifice to pacify the Achesua and Afiada stream gods, as they believe that these pacifications guarantee the community's spiritual safety. But the creeks' past diversion means that the sacrifice now takes place on dry land, where the water used to flow.

Currently, there is no viable recreational activity connected to the rivers and streams in the surveyed area. Apart from isolated instances when children swim in nearby rivers (for example, in the Arura River at Dumasi around May–August), no other water-facilitated recreational activity takes place.

Theme 9: Human rights

In Ghana, the human rights landscape is fairly well established with laws and institutions to promote and protect fundamental rights and freedoms, and/or remedy alleged violations. The Commission for Human Rights and Administrative Justice (CHRAJ) is the national institution responsible for promoting and protecting fundamental rights and freedoms and administrative justice in the country. The commission was established by Act 456 of the 1993 constitution with a three-fold mandate: as a human rights institution; to provide administrative justice (ombudsman responsibility); and as an anti-corruption agency.

The right to water is not explicit in the constitution of Ghana. However, the right to clean water is embedded in the country's constitutional provisions on civil rights and liberties and the International Covenant on Economic, Social and Cultural Rights, which Ghana has ratified. Under the laws of Ghana, the obligations arising out of ratified international covenants have constitutional status (CESR, 2002).

An interview held at the CHRAJ district office at Tarkwa reported that a complaint had been made against Golden Star. There were two parts to this complaint. The first allegation was that leakage and overflow of contaminated water from Golden Star's tailings storage facility had rendered nearby cropland unproductive. The second was that the compensation received from Golden Star for a reclaimed fish pond was not sufficient. An investigation by the Crop Research Institute of the Council for Scientific and Industrial Research found no link between the tailings overflow and the cropland.

On the latter complaint, CHRAJ found that the terms of the agreement for the compensation had been accepted, so it was conducted according to the law.

With regard to restricted access to water, Article 12(2) of Ghana's 1993 constitution stipulates non-discrimination in access to resources and opportunities. Interviewees reported that there is no discrimination with respect to access to water. However, in communities where water facility-user fees are applicable, community members who are economically disenfranchised are unable to pay.

Theme 10: Gender

Typically, the communities in this area are patriarchal, but this does not affect control and access to clean water and sanitation facilities. Generally, access to water and sanitary facilities is non-discriminatory. Both men and women have equal access to communal facilities, except those who are unable to pay user fees. Even then, people who are unable to pay facility-user fees rely on their social support network systems for access. Communal toilet facilities often make equal room for men and women.

At home, women (generally assisted by children) perform most of the daily household chores, including cooking, cleaning, washing and bathing, and are responsible for collecting and storing water for both household and communal use.

With regard to participation in economic activities, men dominate the formal sector of the local economy, including mining and farming. In terms of the informal sector, ASM is considered primarily a male-youth business, while women dominate in agro-processing, food vending, petty trading, and retailing of consumer goods.

At both policy and community levels, the role of women and men is fairly distributed in the management of water facilities. At the district level, the Water and Sanitation Department is responsible for overseeing water facility management. The office purports to be gender-neutral, with a policy of recruiting officials on merit. The CWSA is the main government agency responsible for water policy and planning for small towns and rural communities in Ghana. Regulations require each small town to have a water board, and rural communities to have water and sanitation management (WATSAN) teams, which have the responsibility for managing water and sanitation facilities. Only Prestea and Bondaye have water boards. The rest of the communities, excluding Bogoso, have WATSAN teams. The district planning officer stated that local regulations require 40% of the people in the WATSAN committees to be female, which corroborates the desktop review (Community Water and Sanitation Agency, 2005). The field trip showed that women were represented in the community WATSAN teams; in the Bepo WATSAN, they comprise more than 50% of the representation.

Theme 11: Health

The district planning officer estimated that 80% of the district population has access to clean water. Almost all the respondents attest that everyone has access to water but that more facilities are required for adequate supply. Health officials in this area confirmed that the predominant diseases in the district are malaria, diarrhoea, cholera, upper respiratory tract infections, skin infections, worm infestations, urinary tract infections and anaemia. Out of this list, only upper respiratory and urinary tract infections are not water-related diseases. Malaria is a mosquito-borne disease, but still waters contribute to mosquito breeding. Anaemia is a side effect of frequent malaria attacks and poor nutrition. Health authorities at both the community and district levels suspect that mercury is in the surface water from ASM activities and that contact with such water bodies could give rise to skin irritation. Poor water and sanitation systems may also give rise to diarrhoea, worm infestations and cholera.

Theme 12: Other key stakeholders

The results from the set of questions relating to Theme 12 here have been covered in other themes and will not be repeated.

Theme 13: Interaction between stakeholders

Generally, the site's engagement with local communities is via the Community Consultative Committee. These are grass-roots committees and specific to each of the 16 mine-impacted communities. Its purpose is to provide a platform for discussing community development projects, addressing grievances and sharing information. According to a Golden Star representative, meetings are held monthly and take place in the community. Contrary to this claim, none of the respondents that were interviewed from the community were aware of these meetings, which could indicate that either the meetings were not happening or the community representatives were not giving information back to the broader community.

There are two other committees: the Community Mines Consultative Committee, whose purpose is to discuss matters that are across multiple or all communities and to decide which community projects are funded by the company's development fund; and the Mediation Committee, which at the time of writing has not had to meet because it is the last opportunity for addressing grievances internally before the commencement of legal proceedings.

Apart from these committees, the site also has community information centres manned by community liaison officers, through which aggrieved stakeholders may lodge complaints. The company aims to respond to each complaint within two weeks. Despite all the formal structures to link the company with the community, there is deep-seated mistrust between the site and the community, as multiple interviewees reported that the company had previously failed to respond to their complaints and grievances.

Theme 14: Legislation, policy, politics

The Water Resources Commission Act 1996 established the Water Resources Commission as the sole regulator of water resources, including groundwater and surface water. The primary responsibility of the commission is water resource development and management. Permits are required for water withdrawals (Water Use Regulations, 2001). To extract groundwater, which is a source of water for both the company and the community in this context, licences are needed (Drilling License and Groundwater Development Regulations, 2006). There is a requirement that the water quality be tested and results sent to the commission. The mining company also has to comply with the Environmental Protection Agency Act 1994.

Policies denote the direction and objectives of the government. The National Water Policy (Ministry of Water Resources Works and Housing, 2007) covers both groundwater and surface water. It is an overarching document explaining how the acts and government departments relate to each other.

Whilst not a policy document, the Gender and Water Resources Management Strategy (Water Resources Commission, 2011) has been developed to increase gender considerations in the field of integrated water resource management beyond participation in committees and boards.

The Small Communities Water and Sanitation Policy (CWSA, 2005) is a detailed document on the provision of sanitation and water supply to rural communities. The District Assembly is responsible for the supply of water and sanitation, but the community has input and control of the process via the WATSAN committee. The committee is made up of community members, with women to comprise at least 40% of committee members. The policy document advises that the community should be responsible for a small portion of capital cost (5%) and all of the operation and maintenance. WATSAN is responsible for setting prices. Of relevance in this context are the recommendations in the policy that one borehole should supply no more than 300 people and that people should have access to 20 litres per day with no more than a 500-metre walk to water (Community Water and Sanitation Agency [CWSA], 2005). In general, these service levels are not being met.

Discussion

Many corporate tools for water reporting, such as the World Business Council for Sustainable Development's Global Water Tool and the Global Environment Management Initiative's Local Water Tool, seek to understand business risk related to water from the point of view of the company. The point of difference of the SWAP is that its primary purpose is to survey how the community obtains and uses water, with special attention to the economic and social-cultural characteristics of the context. The results were a summary of the desktop review and the interviews. This discussion section draws out key findings, explains how the results of the SWAP can be used in a practical way by the mining company, and suggests improvements to the tool.

Key findings

Of relevance to this case study is that the company supplies water to supplement inadequate government infrastructure and compensate for diversions, as well as for corporate social investment purposes. For the communities where the mine site has impacted the community's access to surface water due to river diversions, there may be a sense of entitlement to the supplied water. The consequence is that those communities have become dependent on the mining company for water supply and the District Assembly has not had to budget fully for the cost of water supply to smaller rural communities for years. This situation of mining companies acting as surrogate governments often happens in developing countries (Connell, 1997; Garvin, McGee,

Smoyer-Tomic, & Aubynn, 2009), but is not sustainable because there will be a time when the mine site closes. The community and the District Assembly do not appear to be prepared to take control of the water supply.

The SWAP drew out that the primary concern of some communities was whether the groundwater in that area was safe to drink. The researchers did not undertake any monitoring, so it has only been inferred that the oily residue respondents found in the groundwater from some boreholes is from naturally occurring iron-oxidizing bacteria.

Examples of how the mining company in the case study may use SWAP results

When mining companies understand the social water context, they can contribute to the development goals of the communities or regions in which they operate. According to Mutti, Yakovleva, Vazquez-Brust, and Di Marco (2012), developmental corporate social responsibility is where the mining company invests to ensure softer intangible benefits for the community beyond mine closure. A sustainable mining-company-sponsored community development programme is one that can be self-sufficient in the absence of the mining company (ICMM, 2012). For example, had the mining company in this pilot had an understanding of the community's concerns surrounding whether the water was safe to drink, it could have invested in a water monitoring programme, educating people on water quality and the different treatments, along with associated costs. This might then have encouraged more investment by the community in water infrastructure because it appeared that the community was willing and able to pay for water when it was supplied by the local authorities and private suppliers.

Whilst each community in the pilot study had access to water, when water is viewed with a social and cultural lens (Perreault, 2014), it could be argued that the mining company in the context interfered with people's spiritual needs through the river diversion that impacted the Dumasi community. It is hoped that in future, the SWAP results could be used by operational staff to understand the consequences of technical decisions from the community's perspective. The SWAP results are not intended to stay in the community-relations or the social-responsibility domain.

Suggested improvements to the SWAP

The study was a pilot for the SWAP and represented an opportunity to suggest improvements to the tool. The range of themes is comprehensive, and the ability to answer every question would not be found within the company. As was done in the pilot, the company would have to use a combination of desktop research and interviews to obtain the answers.

Desktop research should cover conventional site social/operational/obligatory studies (depending on the country context), including environmental and social impact assessments, socio-economic studies, rapid assessment survey reports and health impact assessments. The data from the desktop sources can be corroborated by interviews.

Gender, socio-cultural and indigenous issues connected to water are better assessed through the collection of primary data – specifically interviews and/or focus group discussions. On average, it took an hour and a half to facilitate a focus group on inter-related themes (see Table 1) in a conversation. The SWAP contains a number of questions where the concepts are repeated across themes, which is useful as different people will give different viewpoints. However, if one respondent is given all the questions to answer, the questions are repetitive, so it would be best to target themes to relevant stakeholders or office holders for interviews.

Table 2 gives suggested data sources – both primary and secondary – for the themes. Table 1 shows the data sources used for the pilot study, while Table 2 provides more general suggestions for stakeholders and data sources and is intended to be more widely applicable.

The pilot identified the following gaps:

(1) In Theme 1, when establishing the geographical reach, no question identifying the communities
(2) No theme on water reporting and disclosure
(3) No theme on legacy issues.

It is necessary to make explicit which communities are included in the study. As stated in the method, the context does not have to be the mine site concession area, although it was in this application. Census data could be used to give the population of the area – disaggregated by gender and age – to help understand potential effects of population pressure on water facilities.

Water monitoring, reporting and disclosure have important roles in effective water management and mitigation of impacts (Wilby & Vaughan, 2011). The disclosure of a particular site's net water consumption, treatment, discharge and water reuse provides evidence to the surrounding community of the potential for impacts, and it has been argued that the community has a right to access site-specific water information (Leong, Hazelton, Taplin, Timms, & Laurence, 2014). The SWAP can be used to populate the

Table 2. Themes of the Social Water Assessment Protocol, linked with possible data sources and stakeholders.

Themes	Possible data sources and stakeholders
Theme 1: Physical water snapshot	Company's water management plan, environmental impact assessment
Theme 2: Climate	
Theme 3: Water supply and infrastructure	Socio-economic assessment reports, environmental and social impact assessments, government reports, interviews with local authorities
Theme 4: Local amenities	Government reports, interviews with local authorities
Theme 5: Domestic uses	Environmental and social impact assessments, government reports, interviews with local authorities and domestic users
Theme 6: Formal and informal economy	Rapid assessment reports, environmental and social impact assessments, national surveys, interviews with stakeholders from livelihood sectors
Theme 7: Indigenous peoples	Interviews with members of indigenous groups and community members
Theme 8: Social, cultural and spiritual	
Theme 9: Human rights	Human rights commission
Theme 10: Gender	Interviews with community
Theme 11: Health	Health impact assessments, government reports, interviews with local authorities
Theme 12: Other key stakeholders	
Theme 13: Interaction between stakeholders	Site stakeholder engagement strategy/register, site stakeholder mapping report, grievance management procedure, company reports, interviews with stakeholders
Theme 14: Legislation, policy, Politics	Government acts, policies, regulations

Contextual Statement of the Minerals Council of Australia Water Accounting Framework, or can supplement other corporate water reports, and so there should be a theme around identifying the standards or initiatives that the company reports to and examining the potential for integration.

There is a need for the SWAP to include questions on the history of mining in the context, mine ownership and previous companies' water management policies – that is, legacy issues – in order to appreciate present-day concerns. Interviews with respondents in Prestea were dominated by legacy issues; however, they did not relate to water and so are not present in the report.

The following points are aimed at refining the SWAP questions. Questions that ask for percentages are difficult for people to answer in an interview format because they require knowledge of statistics. If a percentage must be estimated, then the tool could suggest that this information be drawn from reports or census data rather than a prompt question in an interview. If the question must be asked during an interview, then it should be modified to ask the respondent for quantities or amounts.

A question in Theme 11, Health, asks about "water-specific community health issues". It is suggested that the user of the SWAP tool ask a broader question first to determine the common diseases or health issues, and then narrow down to water-related ones, because the question as it stands does not reveal whether water-related health issues are significant in the region. Similarly, there should be a broader question under Theme 6, Formal and Informal Economy, to first determine the major livelihoods of the context and then ask about those which use water to provide a better understanding of the structure of the economy. The questions within Theme 6 distinguished between between formal and informal economies, but not one interviewee distinguished between the two types when providing answers.

In summary, the themes and questions themselves require only minor revision, but the usability of the SWAP questions would be improved if questions could be segregated within the themes by those that are to be answered from desktop data sources versus those that are to be answered by interviews and that questions are framed accordingly.

Conclusion

The application of the series of questions in the Social Water Assessment Protocol showed that the tool was an effective and systematic way to draw out the social issues related to water in the communities that surround the Bogoso/Prestea Golden Star mine site concession. The outcomes of the pilot were that it refined some questions and suggested the addition of some themes, but most importantly, the pilot created a table identifying the data sources and stakeholder groups that could best provide the information required for the SWAP.

Potential users of the SWAP are a mine site's social responsibility and community relations teams, and the operational or environmental staff who manage water. For the first group, the SWAP can assist in developing corporate social responsibility or community development projects which will benefit the community after mine closure. For the mine water management staff, knowing the social implications of water management options may encourage more culturally sensitive solutions.

Notes

1. http://www.getamap.net/maps/ghana/ghana_(general)/_bogoso/
2. http://population.mongabay.com/population/ghana/2295840/prestea

Acknowledgements

The authors acknowledge the contribution of Justice Oppong Mensah, a local community member who acted as an interpreter.

Funding

This project was funded by the International Mining for Development Centre. The centre was established as a joint venture between the University of Western Australia and the University of Queensland, with grant funding from the Australian Government through an Australian Aid initiative. The authors disclose that Golden Star Resources provided food and accommodations during the field trip for two members of the research team.

References

Anglo American. (2007). SEAT: Socio-Economic Assessment Toolbox. Version 2 Retrieved 27th October, 2014, from http://www.angloamerican.com.au/development/social/community-engagement/~/media/Files/A/Anglo-American-Plc/siteware/docs/seat_toolbox2.pdf

CESR (Centre for Economic and Social Rights). (2002). *Report of the International Fact-Finding Mission on water sector reform in Ghana*, Retrieved 14th November 2013 from http://cesr.org/downloads/factfindingmissionGhana.pdf

CHRAJ (Commission on Human Rights and Administrative Justice). (2008). *The State of Human Rights in Mining Communities in Ghana*. Accra, Ghana: Author. Retrieved from http://www.ircwash.org/resources/small-communities-water-and-sanitation-policy.

Collins, N., & Woodley, A. (2013). Social water assessment protocol: A step towards connecting mining, water and human rights. *Impact Assessment and Project Appraisal, 31*(2), 158–167. doi:10.1080/14615517.2013.774717

Community Water and Sanitation Agency. (2005). *Small Communities Water and Sanitation Policy*. Accra, Ghana: Community Water and Sanitation Agency. Retrieved from http://www.ircwash.org/resources/small-communities-water-and-sanitation-policy

Connell, J. (1997). *Papua New Guinea: The struggle for development*. Studies in the Growth Economies of Asia Series. London: Routledge.

Danoucaras, N., Babatu, A., & Sturman, K. (2014). Participatory Water Monitoring Scoping Study and SWAP Pilot in Ghana. Retrieved 28th October, 2014, from http://im4dc.org/wp-content/uploads/2014/10/Danoucaras_PWM-Completed-Report.pdf

Dupuy, K. E. (2014). Community development requirements in mining laws. *The Extractive Industries and Society, 1*(2), 200–215. doi:10.1016/j.exis.2014.04.007

Franks, D. M., & Vanclay, F. (2013). Social Impact Management Plans: Innovation in corporate and public policy. *Environmental Impact Assessment Review, 43*, 40–48. doi:10.1016/j.eiar.2013.05.004

Garvin, T., McGee, T. K., Smoyer-Tomic, K. E., & Aubynn, E. A. (2009). Community–company relations in gold mining in Ghana. *Journal of Environmental Management, 90*(1), 571–586. doi:10.1016/j.jenvman.2007.12.014

GEMI. (no date). Collecting the drops: A water sustainability planner. Retrieved 27th October, 2014, from http://www.gemi.org/waterplanner/Documents/GEMI-Water-Mgt-Risk-Assessment-Questionnaire.pdf

Ghana Statistical Service (GSS). (2014). *2010 Population and Housing Census: District Analytical Report*. Prestea/Huni Valley District. Accra: Ghana: Author.

Goff, M., & Crow, B. (2014). What is water equity? The unfortunate consequences of a global focus on 'drinking water'. *Water International*, *39*(2), 159–171. doi:10.1080/02508060.2014.886355

Golden Star Resources. (2013). Historical mining operations at Bogoso/Prestea. Retrieved 18th December, 2013, from www.gsr.comOperationsBogoso.asp

Hilson, G., & Yakovleva, N. (2007). Strained relations: A critical analysis of the mining conflict in Prestea, Ghana. *Political Geography*, *26*, 98–119. doi:10.1016/j.polgeo.2006.09.001

ICMM (International Council on Mining and Metals). (2012). Community Development Toolkit. Retrieved 21st August, 2015, from http://www.icmm.com/document/4080

Kemp, D., Bond, C. J., Franks, D. M., & Cote, C. (2010). Mining, water and human rights: Making the connection. *Journal of Cleaner Production*, *18*(15), 1553–1562. doi:10.1016/j.jclepro.2010.06.008

Leong, S., Hazelton, J., Taplin, R., Timms, W., & Laurence, D. (2014). Mine site-level water reporting in the Macquarie and Lachlan catchments: A study of voluntary and mandatory disclosures and their value for community decision-making. *Journal of Cleaner Production*, *84* (0), 94–106. doi:10.1016/j.jclepro.2014.01.021

MCA (Minerals Council of Australia). (2012). Water Accounting Framework for the Minerals Industry User Guide Version 1.2 - April 2012. Retrieved 20 June, 2013, from http://www.minerals.org.au/focus/sustainable_development/water_accounting

McNamara, N. (2013). Corporate social responsibility and compliance: Transnational mining corporations in Tanzania. *Macquarie Journal of International and Comparative Environmental Law*, *9*(2), 1–17.

Ministry of Water Resources Works and Housing. (2007). *National Water Policy*. Accra, Ghana: Author.

MMSD (Mining, Minerals and Sustainable Development). (2002). Breaking New Ground. Retrieved August 21, 2015, from http://pubs.iied.org/pdfs/G00901.pdf

Mutti, D., Yakovleva, N., Vazquez-Brust, D., & Di Marco, M. H. (2012). Corporate social responsibility in the mining industry: Perspectives from stakeholder groups in Argentina. *Resources Policy*, *37*(2), 212–222. doi:10.1016/j.resourpol.2011.05.001

National Ground Water Association. (2009). Iron Bacteria. Retrieved 25th July, 2014, from http://www.ngwa.org/Documents/ClipCopy/Iron-Bacteria.pdf

Perreault, T. (2014). What kind of governance for what kind of equity? Towards a theorization of justice in water governance. *Water International*, *39*(2), 233–245. doi:10.1080/02508060.2014.886843

Sethi, S. P., Lowry, D. B., Veral, E. A., Shapiro, H. J., & Emelianova, O. (2011). Freeport-McMoRan Copper & Gold, Inc.: An Innovative Voluntary Code of Conduct to Protect Human Rights, Create Employment Opportunities, and Economic Development of the Indigenous People. *Journal of Business Ethics*, *103*(1), 1–30. doi:10.1007/s10551-011-0847-4

UNGA (United Nations General Assembly). (2010). Resolution on the human right to water and sanitation. UN doc. A/RES/64/292, from http://www.un.org/en/ga/search/view_doc.asp?symbol=A/RES/64/292

Water Resources Commission. (2011). *Gender and Water Resources Management Strategy*. Accra, Ghana: Author. Retrieved from www.wrc-gh.org/dmsdocument/22d

Wilby, R. L., & Vaughan, K. (2011). Hallmarks of organisations that are adapting to climate change. *Water and Environment Journal*, *25*(2), 271–281. doi:10.1111/wej.2011.25.issue-2

Unconventional oil and gas extraction in South Africa: water linkages within the population–environment–development nexus and its policy implications

Surina Esterhuyse, Nola Redelinghuys and Marthie Kemp

ABSTRACT

The development of unconventional oil and gas resources, controversial in many countries, is currently being pursued by the South African government. This activity can have large impacts on the socio-economic and biophysical environments, especially water resources. In South Africa, little consideration has been given to water-related impacts from the perspective of the inter-related people–ecosystem linkages that are necessary for sustainable social and economic development. This article explores specific water-related linkages between facets of the natural and social environments pertaining to unconventional oil and gas extraction, with the objective of achieving more effective water resources management and water policy development.

Introduction

The growing demand for energy challenges economic development in South Africa. The South African population is predicted to grow from 50.6 million currently to 58.5 million by 2030. Migration from neighbouring countries could push this number up to 61.5 million (Republic of South Africa. National Planning Commission [RSA NPC], 2013), driving the need for energy, jobs and food security. In 2012, approximately 85% of South African households had access to electricity for domestic use, but electricity supply services have declined nationally from previous years, with 17.5% of households reporting electrical interruptions in 2012 (Statistics South Africa, 2013). By 2030, at least 29,000 MW of additional electricity will be required to cater to South Africa's growing energy needs; and with 10,900 MW of capacity due to be retired by then, new builds of more than 40,000 MW will be required (RSA NPC, 2013). In light of this, the government has decided to proceed with the exploration and extraction of unconventional oil and gas (UOG) resources (RSA NPC, 2013). Another concern driving the current focus on UOG is carbon emissions. Of the world's countries, South Africa is the twelfth-largest carbon emitter (RSA NPC, 2013), because its energy-intensive economy is currently heavily dependent on carbon-based fuels. UOG is viewed as a transitional fuel that could, in the long term, assist in reducing carbon emissions because shale gas has a lower carbon footprint than coal, producing less greenhouse gas emissions per MWh of electricity than coal (Cohen & Winkler, 2014; Wait & Rossouw, 2014). The first exploration licences have been awarded in South Africa (PASA, 2015) and exploratory drilling could commence, depending on timeframes for obtaining environmental authorizations, while full-scale extraction of the resources may take longer and would depend on the economic viability of extracting the resources.

In addition to the above energy constraints, South Africa is also a water-stressed country, placing additional challenges on future economic development. The country receives an average rainfall of 497 mm per year, well below the global terrestrial average of 860 mm, and has annual freshwater availability of approximately 1000 m^3 per person per year (Claassen, 2010). Water availability is becoming a serious constraint on future social and economic development, as a high percentage of available water resources has already been allocated (Department of Water Affairs [RSA DWA], 2013). UOG extraction may place additional pressures on these already over-allocated water resources.

UOG extraction and the management of its related impacts are controversial in many countries worldwide, such as Canada, France, the US and the UK. This relatively new oil and gas production technique – essentially a mining technique (Binnion, 2012) – requires the extraction of these resources by enhancing the permeability of the target oil or gas reservoir. Stimulation is usually achieved by hydraulic fracturing, where a high volume of water, proppant and chemicals is injected into the reservoir to enhance reservoir permeability and extract the oil or gas resources from the reservoir, although other stimulation methods such as acidizing or depressurization may also be used (Broomfield, 2012).

The processes by which UOG resources are extracted are – from environmental and social perspectives – controversial. Those who favour the development of UOG resources emphasize that these developments will contribute to energy provision and enhance socio-economic well-being (Brasier et al., 2011; Wait & Rossouw, 2014; Warren, 2013). Those who are opposed emphasize that socio-economic benefits will not be equally distributed, and that the developments will result in untold negative impacts on the social and biophysical environments (Schafft, Borlu, & Glenna, 2013; Warren, 2013).

Any of the activities associated with UOG extraction might seriously impact both the biophysical and the socio-economic environments. For example, while oil and gas exploration and extraction may enhance socio-economic development in certain areas through increased primary and secondary employment opportunities, environmental impacts such as depletion of water sources, water contamination and seismicity may impact on community health and safety, and food security for the poorer sectors of these communities. Transdisciplinary studies on the linkages between the socio-economic and biophysical impacts of UOG in South Africa are required to ensure that sustainable human development is not impeded. This is especially necessary while UOG development is still in its infancy there.

The article firstly provides a background to the South African policy framework that emerged from a growing recognition of the importance of including population, environment and development issues in decision making surrounding environmental issues. It then offers a discussion of the population–environment–development nexus (PED-nexus), the framework of analysis used in this article.

The PED-nexus forms the basis for illustrating the linkages between UOG extraction-related impacts within the nexus, with a specific emphasis on water-related linkages. Finally, it discusses the implications of these linkages within the South African water policy framework, identifies policy shortcomings and proposes possible solutions.

Background to the South African policy framework

The South African environmental policy framework recognizes several international conventions and agreements on sustainable development, including the 1992 Rio Declaration on Environment and Development, the 1994 International Conference on Population and Development's Programme of Action and the 2002 World Summit on Sustainable Development (Brynard & Stone, 2004). In their *Progress Review of the Implementation of the White Paper on Population Policy for South Africa*, the Department of Social Development (RSA DSD, 2010) mention that these ideals are entrenched in, among others, the Constitution of the Republic of South Africa, Act 108 of 1996 (RSA, 1996), the Population Policy of South Africa (1998) and the South African National Environmental Management Act (NEMA), Act 107 of 1998 (RSA, 1998a). The Constitution and the NEMA act as the framework within which South Africa's environmental legislation should operate. According to the Constitution, every person has the right to have the environment protected through legislative and other measures that prevent pollution and degradation. These measures must secure the sustainable use of natural resources while promoting justifiable economic and social development. The NEMA acknowledges the interdependence of socio-economic and biophysical systems. One of the key principles of the NEMA requires that all development in South Africa be socially, economically and environmentally sustainable. The NEMA defines sustainable development as the "integration of social, economic and environmental factors into planning, implementation and decision-making so as to ensure that development serves present and future generations".

In spite of the ideal of sustainable development, approaches to policy development and implementation in South Africa remain fragmented – both laterally, between different government departments, and hierarchically, between national, provincial and local levels of government. This fragmentation is acute in the development of policy related to UOG extraction in the country (Centre for Environmental Rights [CER], 2013). This article describes the fragmented way in which the linkages between social and biophysical factors are accounted for within the current policy framework.

Considering that the South African government places a high priority on sustainable human development, it is imperative that a potentially harmful activity such as UOG extraction be studied in a manner that recognizes the interrelated nature of potential impacts. A thorough understanding of the interconnectedness of water-related impacts, socio-economic development and the human population can support and strengthen integrated policy development and planning with regard to UOG extraction. Such an analysis can also offer insights into the complexity surrounding UOG extraction with regard to water-related impacts and the linkages thereof with socio-economic development and population issues.

Conceptualizing interactions between people and ecosystems

The complex linkages between the biophysical and social environments can be moulded into an analytical systems framework known as the population–environment–development nexus (Hummel et al., 2013). The PED-nexus expresses the complex and reciprocal interface between population, environment and development factors, with these factors being in constant and dynamic interaction with each other (Groenewald, 2011; Pelser & Redelinghuys, 2008). The relationships between factors are constantly subject to redefinition as the dynamic interaction changes the constitution of the relationships in the nexus.

Population includes the size and composition of a population, and the processes of migration, fertility and mortality.

Environment encapsulates the functions necessary for human survival, well-being and socio-economic development, namely to provide a living space for people, act as a resource repository (for instance by providing drinking water), and act as a sink for the end products and waste of human consumption (for instance, as a receptacle for the waste generated in the extraction of UOG).

Development refers to industrial, agricultural and infrastructure activities. Development should be sustainable, meeting the current needs of the population in a way that the ability of future generations to meet their own needs is not compromised (O'Riordan, 2007).

Implicit in the PED-nexus framework is the idea of environmental limits and boundaries, where resources are depleted if the pace of resource consumption exceeds the rate of regeneration. The ability of the biophysical environment to act as a waste repository is strained when human waste production exceeds the ability of the environment to absorb the waste created, negatively impacting human health and well-being, quality of life, and sustainability. The PED-nexus is illustrated in Figure 1.

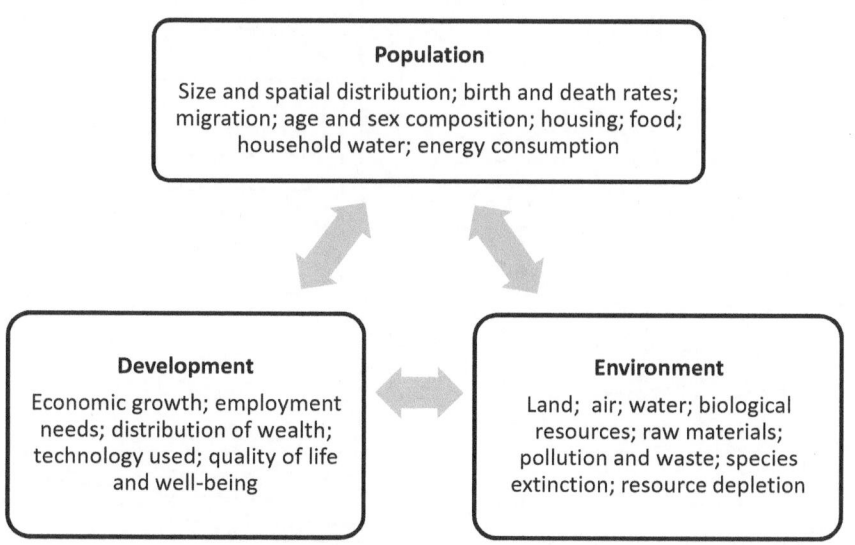

Figure 1. The population–environment–development nexus.
Source: Pelser and Redelinghuys (2008).

The PED-nexus framework forms the basis for analyzing international and national policy frameworks (Hummel et al., 2013; Pelser & Redelinghuys, 2008). As will be demonstrated, it is especially suitable for analyzing the complex linkages presented by UOG extraction – an issue that has far-reaching implications ranging from the macro level (national energy provision) to the micro scale of community health and well-being.

UOG extraction and water linkages within the PED-nexus

This article analyzes the interplay between water-related environmental factors (groundwater and surface water quantity and quality), and the linkages of these factors with various population and development factors. Focusing on these specific impacts is not meant to suggest that they constitute the entire sum of possible linked impacts pertaining to UOG extraction; however, in view of the specific environmental and socio-economic context of South Africa, in which issues such as water scarcity, food insecurity and poverty are paramount, these issues were selected to demonstrate the linkages between different aspects in the PED-nexus. It should also be noted that the identified impacts associated with UOG extraction that are used to illustrate linkages within the PED-nexus are based on the international literature and that what happens in one country might not necessarily happen in another country.

Figure 2 presents a simplified representation of the complexity of the interactions and linkages between the different water-related environmental, developmental and

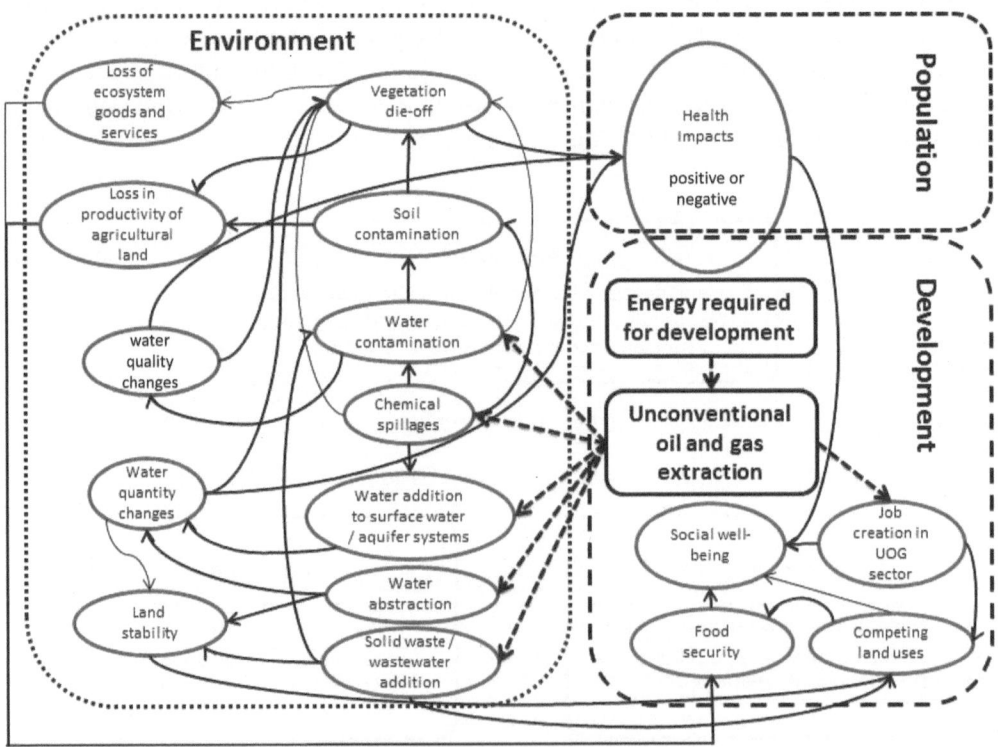

Figure 2. Representation of linkages between unconventional oil and gas extraction water-related impacts within the population–environment–development nexus.

population aspects of UOG extraction within the PED-nexus framework. Although the impacts described here are mostly negative, we recognize that some are not. Figure 2 illustrates that development drives energy requirements and, in this case, UOG extraction. Various water-related impacts in the environmental sphere of the nexus can be linked directly with UOG extraction (dashed arrows), including above-ground chemical spillages, leading to soil contamination and changes in water quality and quantity; direct water pollution through migration of chemicals in aquifer systems; the production of solid waste; water abstraction during coalbed methane extraction (Australian National University [ANU], 2012); and water addition (during fracking operations). These impacts are linked to impacts on other environmental entities, for instance water quality and quantity changes that can cause vegetation die-off (Adams, 2011), which in turn, together with soil contamination, can lead to a loss of productivity in agricultural land. This loss of productivity can adversely affect the social well-being of farming communities and, over time, negatively impact food security. Water-related aspects such as water sourcing, wastewater disposal and poor solid waste management can impact the quantity and quality of water available in aquifers and river systems and can also induce geological instability and aquifer connectivity due to deep underground injection of wastewater (ANU, 2012). In the population sphere of the nexus, UOG extraction can contribute to a decline in the health of populations as a result of environmental contamination arising from UOG extraction. This can be in the form of an increased risk of cancers and organ damage and the worsening of chronic conditions due to water quantity and quality changes, especially in vulnerable populations like children and the elderly (Broderick et al., 2011; Considine, Watson, & Considine, 2011; Dolesh, 2011).

The subsequent discussion starts from the premise that that energy is required for development. It then explores the energy–water linkages, water–agricultural linkages and water–population linkages of UOG extraction in South Africa. An understanding of these could aid in the development of an integrated policy framework in which water resources are protected and sustainable socio-economic well-being is promoted.

Energy development and the demand for water

According to the Karoo shale gas report (Econometrix, 2012), the energy requirements for South Africa's middle and high economic growth scenarios (4.5% and 7% per year, respectively) will be dependent upon the expansion of the available energy base. The South African economy is highly dependent upon electricity production for its industrial, commercial and domestic energy needs. Coal, with an 81% share of energy generation capacity, is the primary energy source utilized by the parastatal energy utility company Eskom, followed by nuclear (4.56%), pumped storage (3.55%), hydro (1.5%) and gas turbines (0.86%). Power generation capacity outside of Eskom is 6.0% by municipalities and 2.0% by private companies (Econometrix, 2012). If UOG proves to be physically and economically extractable in South Africa, it will augment primary energy supplies, replace energy imports, facilitate the transition from nonrenewable to renewable energy sources and also boost the economy (Econometrix, 2012; RSA NPC, 2013). These positive impacts would, however, require infrastructure to be developed to capture, process and transport the oil and gas to the end user. It would also be necessary to ensure proper operation and maintenance of processing plants and implement good environmental practices such as wastewater recycling and waste minimization.

Factors that may hamper the development of UOG in South Africa include a low international oil price (Cooke, 2015), the economic viability of extraction (Binnion, 2012) and water availability (Hedden & Cilliers, 2014). Water availability may severely constrain the development of most power generation options in South Africa. The low average rainfall and high evaporation rate, for example, contribute to a lack of perennial rivers necessary for hydroelectric power generation, while the development of UOG resources – specifically shale oil and gas – may require large amounts of water during extraction. Water requirements for each shale production well may range widely, with reported usage of between 10,000 m^3 and 30,000 m^3 of water for the fracking procedure (Grant & Chrisholm, 2014), excluding ancillary water requirements such as for sanitation and dust suppression. Water requirements for coalbed methane are generally lower than for shale oil and gas operations (USEPA, 2011), but wastewater management becomes more challenging (ANU, 2012).

There is, therefore, an inextricable link between water availability and energy resource development. Water is required for the development of energy, but energy is also required to secure, deliver, treat and distribute water (Scott et al., 2011; Siddiqi & Anadon, 2011). The linkage between water and energy requires a tandem approach to the management of these resources by government and other stakeholders in the water sector, on both local and national scales. Yet, despite these linkages, there are few examples of the tandem management of water and energy resources internationally (Electric Power Research Institute [EPRI], 2002a, 2002b; Sovacool & Sovacool, 2009), and in South Africa such tandem management must still be achieved.

Water, agricultural livelihoods and food security

As a competing water and land use to agriculture in South Africa, the development of UOG resources may endanger the preservation of local agricultural livelihoods. Whilst commercial farmers may benefit from the sale of their land to UOG extraction companies, their farm workers may become unemployed. In South Africa, agricultural employment already declined by 30.9% between 2001–2014 as farmers abandoned farming for other enterprises and 'labourless' technology was introduced (South African Institute of Race Relations [SAIRR], 2014). Internationally, UOG extraction results in abandonment of farms and a negative impact on employment in agriculture and related industries through its effect on the water supply (De Rijke, 2013). The expected conversion of land use from agriculture to UOG extraction can also exacerbate food insecurity (Millar & Roots, 2012).

Agriculture remains important to South Africans, with 10.5 million hectares of land being used for commercial agricultural purposes (United Nations Department of Economic and Social Affairs, 2007) and almost 62% of South Africa's water resources being allocated to commercial agriculture (Department of Water Affairs [RSA DWA], 2009). Approximately one million people are employed directly by the agriculture sector, and some six million South Africans depend on it indirectly for employment.

UOG extraction might coexist with other forms of agriculture, such as livestock farming, where land is used primarily for grazing rather than commercial crop production. UOG extraction is also a potential employment creator in rural areas, where job opportunities in the agricultural sector have been declining (Brasier et al., 2011). A report by Econometrix (2012) tested four scenarios, with consumed gas resource sizes ranging between 0.44% and 4.44% of the then estimated 450 trillion cubic feet of gas. Total upstream and downstream employment in the lowest UOG development scenario (0.44% of desktop reserve estimates) was approximately 48,070 jobs, while the highest resource scenario tested (4.4% of desktop reserve estimates) indicated peak total employment of 355,817 jobs. The expected oil and gas resource was, however, greatly reduced to approximately 30 trillion cubic feet later in 2012. Furthermore, job creation in the UOG sector may come at the expense of current employment in agriculture (Beemster & Beemster, 2011). While some skilled workers may be absorbed by the UOG sector and secondary commercial enterprises, unskilled and semi-skilled farm workers may suffer a different fate. Those who become disconnected from their livelihoods in rural areas usually migrate to urban areas in search of economic opportunities, increasing the number of destitute migrants in impoverished urban areas (Pelser & Redelinghuys, 2008) and putting additional pressure on water delivery and sanitation services in urban areas (Dondeynaz, Moreno, & Lorente, 2012).

UOG extraction, water quantity and quality, and human populations

In its effects on both surface water and groundwater quality and quantity, sourcing of water for fracturing operations could impact water availability. In South Africa, some 300 towns and over 65% of the population are fully reliant on groundwater sources for domestic uses (Woodford, Rosewarne, & Girman, n.d.). These are located in the Karoo Basin, in the central part of the country, where UOG extraction is planned (De Wit, 2011).

At least 10% of deaths in children under five can be attributed to a lack of access to safe water, while 84% of all deaths from diarrhoea in South Africa are linked to a lack of access to safe water and adequate sanitation (Harrison, 2010; Redelinghuys, 2012). Given these figures, the additional impact of UOG extraction on water quality and quantity is concerning.

Water requirements for UOG extraction are directly proportional to the density of gas wells. Oil and gas well densities can range upwards from one well per km^2 (IEA, 2012). Each well requires between 10,000 m^3 and 30,000 m^3 of water per fracturing operation (Broomfield, 2012; Grant & Chrisholm, 2014). If this water is abstracted from groundwater, it could affect aquifer connectivity and ultimately deform aquifers (ANU, 2012; Williams, Stubbs, & Milligan, 2012). Abstraction from surface water also impacts human uses such as crop production and domestic water use (Anderson & Theodori, 2009).

UOG extraction and its related activities (vegetation clearing for well pad construction, access roads, and pipelines) may furthermore have direct impacts on vegetation, with an associated loss of biodiversity. Degraded ecosystems are unable to persist in the face of the extensive ecological footprints of gas industries, especially when the impact is compounded by factors such as climate change (Baynard, 2011). The provision of ecosystem goods and services may also significantly decrease in South Africa's low-rainfall, arid environment (Milton, 2012). The loss of ecosystem services may pose challenges for subsistence-based populations who specifically rely directly on natural resources such as wood, grass, game, and edible and medicinal plants for income and subsistence (Pelser, 2012). Contamination of surface water resources through spills can also affect the habitat of aquatic species like fish that local communities rely on as wild food sources (Walsh, 2011).

Lastly, solid waste and wastewater disposal practices can adversely affect water resources (Vengosh, Jackson, Warner, Carrah, & Kondash, 2014), and eventually people. Brines generated during wastewater treatment need to be disposed of (ANU, 2012), and if wastewater is injected into deeper porous geological formations, it may cause geological and aquifer deformation, which may trigger fluid migration and water contamination (Broomfield, 2012), or seismicity, with associated land instability issues (Lechtenböhmer et al., 2011; National Research Council [NRC], 2012; Zoback, Kitasei, & Copithorne, 2010). Poor management of solid waste and wastewater may also sterilize land for other uses (Broomfield, 2012; IEA, 2012), and if it contaminates groundwater resources, the complex and heterogeneous aquifers in South Africa may be very difficult, if not impossible, to remediate. The NRC (2012) states that restoration of groundwater resources contaminated by anthropogenic releases remains a significant technical and institutional challenge in the US and that at more than 126,000 sites across the US, groundwater contamination occurs at such levels that closure for these sites could not be obtained. For a developing country such as South Africa, wastewater management and water resource protection may present great challenges, which need to be dealt with through the development and implementation of appropriate policies and legislation.

Scale of impacts and water policy implications

Scale of impacts

The scale of the impacts of UOG extraction spans both spatial and temporal dimensions and is cumulative. On a spatial scale, UOG extraction does not occur only within specific geographic boundaries, as is the case for localized mining operations, but includes an array of gas well sites that may cover vast geographic expanses (ANU, 2012; IEA, 2012; Williams et al., 2012). In South Africa, current oil and gas exploration applications cover almost half of the surface area of the country (PASA, 2015). Apart from the specific geographic location where oil and gas are extracted, a wider expanse of surface area is usually impacted by the connecting roads, processing plants and pipelines that serve to process the product and transport it to the end user (Broomfield, 2012; IEA, 2012). Therefore, environmental and socio-economic impacts may cross provincial or other boundaries, such as catchments or municipal limits, in ways that localized mining operations usually do not (Holahan & Arnold, 2013). This may complicate efforts to coordinate integrated resource management (Holahan & Arnold, 2013) to protect resources such as water or livelihoods. It also makes the development of UOG extraction regulatory policy for water and other resources quite complex (Warner & Shapiro, 2013).

In some instances, UOG extraction impacts can even cross international political borders. A case in point is UOG extraction in Botswana, one of South Africa's neighbouring countries. This may impact the south-eastern Kalahari/Karoo Basin transboundary aquifer, which spans South Africa, Namibia and Botswana (Vasak, 2008), while UOG extraction in the Kgalakgadi National Park may endanger biodiversity on the South African side of the park (Barbee, Dutschke, & Smith, 2013). Cross-border impacts can either drive conflict between regions (Rabe & Borick, 2013) or countries, with such cases testing the strength of current transboundary agreements governing the protection and use of shared natural resources – or they could open an avenue for more cooperation between neighbouring countries.

There is also a spatial dislocation between the UOG that is to be extracted and the water required to perform this extraction. Oil and gas shale deposits occur in the Karoo geological basin, a very arid area with a high dependence on limited groundwater resources (Woodford et al., n.d.). Extracting the UOG from this basin may demand resource coupling and management beyond the point of use of either resource if the groundwater resources in this region prove to be insufficient. Resource coupling could include the transportation of water from other areas, with associated water management challenges involving multiple jurisdictions.

The most important negative externalities associated with UOG extraction include economic wastes and environmental impacts. Pinpointing the precise locations for optimal production of oil or gas from shale is often difficult due to technological limitations and geological uncertainty and heterogeneity. Operators would typically perform extensive drilling and fracking operations to create an expansive fracture network and optimize possible production from shale resources (Holahan & Arnold, 2013). Due to the possible complex and heterogeneous nature of the geology in oil or gas well fields and the expansive fracture networks that may have been created by operators, fracking fluid or contaminated groundwater may migrate via artificial fracture systems to aquifers beyond the surface extent of a single gas well or a gas well field. The geology of individual shale gas plays may also differ significantly (IEA, 2012). Because the underground source of these environmental impacts may go unnoticed, and because these impacts represent non-point pollution sources, operators do not have an incentive to internalize these impacts (Holahan & Arnold, 2013). Consequently, authorities may experience difficulties in delineating the extent around an oil or gas well or demarcating related areas where water resources should be monitored and protected. Assigning liability in the case of contamination also becomes difficult.

The impacts of UOG extraction are cumulative, and such impacts do not scale linearly, because ecosystems often respond to perturbations in non-linear and often unpredictable ways (Scheffer et al., 2012). These non-linearities are often not considered in strategic management of activities that impact on the environment (Buschke & Vanschoenwinkel, 2014).

Impacts can persist for much longer than the life of the gas production enterprise. Groundwater contamination due to oil and gas well infrastructure that fails over time may only be identified much later, after UOG extraction in a specific area has ceased. A large variation in percentage of well leakages over time, space and type of well has been reported in the US, with influencing factors being well integrity, local geology and the extent of fracture systems that provide pathways for contamination. Groundwater and soil contamination through disposal of waste or wastewater containing naturally occurring radioactive material can also leave an environmental legacy that builds up over time (Vengosh et al., 2014) and could impact local wildlife and communities.

Lastly, institutional scales also play a role in the management of UOG extraction. Nationally, in South Africa, UOG is portrayed as critical to South Africa's energy independence, and as having lighter environmental impacts than the coal industry that currently supplies the largest part of South Africa's energy (Econometrix, 2012). However, controversies exist over UOG development, characterized by different decision-making issues at different institutional levels and competing goals, notably a higher economic gain at the national level than at the local level, but with environmental externalities that must be handled at the local level. Further examples of controversies include uncertainties regarding the amount and location of water required for UOG extraction, the composition of fracking fluids, and the extent of impacts related to UOG extraction. Addressing limited time and poor institutional resources in South Africa is critical for successful water resource management and protection during UOG extraction (CER, 2013). The South African National Water Resource Strategy specifically identified inadequate financial resources, human resources and skills shortages, as well as limited institutional capacity, as critical factors that must receive priority attention for the effective management of water resources (RSA DWA, 2013).

Policy implications

Resources are typically managed at multiple institutional scales. In South Africa these scales can be divided into local (municipal and provincial administrations), national (government departments), and for UOG resources, international (e.g. with Botswana). At the local level, different municipalities and provinces in South Africa would need to coordinate their management to ensure protection of shared water resources during UOG extraction. On the national level, policy making must extend to take sovereignty concerns into account.

Local concerns over water-related environmental and human health impacts need to be considered within policy frameworks and should not be handled as engineering problems. Although local decision makers may have limited capacity or authority to address such impacts, they must be assisted through regulatory cooperation with higher institutional scales. Enforcement of regulations must also be strengthened by providing adequate human and financial resources. The range of possible water-related impacts, and the fact that these impacts interlink in the PED-nexus, also requires multi-tiered institutional management of water, oil and gas, and other resources. Hence, government departments with different mandates would need to cooperate if UOG extraction and water resource protection were to be managed responsibly in South Africa.

Multi-tiered institutional arrangements with laws, policies and organizations that operate across administrative boundaries to manage resources often offer a wider set of management options (Scott & Pasqualetti, 2010). These can be achieved through integrated assessment (using strategic environmental assessments in addition to environmental impact assessments), through planning that takes into account all the resources that may be impacted upon during UOG extraction, and by identifying key trade-offs that cross sectors, resources and boundaries (Hightower & Pierce, 2008).

When identifying trade-offs, the energy development portfolio for the country must take into account resource limitations (Department of Energy [USA DOE], 2006). In arid countries with high climatic variability, such as South Africa, using less water for energy development may override other concerns, such as compliance with carbon emission standards.

Since UOG extraction has the potential for diffuse and widespread negative environmental impacts, especially water contamination, for which assigning liability could be difficult, Holahan and Arnold (2013) argue that the ideal regulatory policy is one that employs the precautionary principle, which is highly supported in South Africa. Regulators have been drafting relevant regulations under existing legislation to minimize water-related environmental impacts linked to UOG extraction (CER, 2013). To date, draft technical regulations on the exploration and exploitation of petroleum resources have been published by the South African Department of Mineral Resources under the Mineral and Petroleum Resources Development Act, Act 28 of 2002 (RSA, 2002) in October 2013. The Department of Water Affairs (now called the Department of Water and Sanitation) published a notice of intent to declare the "exploration for and or production of onshore unconventional oil or gas resources and any activities incidental thereto including but not limited to hydraulic fracturing" a controlled activity under the National Water Act, Act 36 of 1998 (RSA, 1998b) in August 2013. The department is currently drafting regulations to protect water resources during UOG extraction. Although regulations are being developed in South Africa to manage UOG extraction, the implementation of these regulations is paramount for the effective management of UOG extraction and ultimate protection of the environment.

In regulations for the protection of water resources, regulators could specify setback distances between UOG wells and important water resource points and areas such as boreholes, water reservoirs, and rivers in order to protect water resources from contamination. They may also require that UOG wells be spaced such that their distance reflects the larger areal extent of horizontal wells and their fractures, which could make assignment of liability for water-related environmental impacts easier. The non-point source nature of environmental impacts also makes it extremely important to employ water quantity and quality monitoring throughout all the UOG extraction phases, from creating a baseline *ex ante*, through the extraction period, to after it is finished (Holahan & Arnold, 2013). Surveys indicate strong support among decision makers for extensive monitoring of water resources in South Africa (Esterhuyse, Kemp, & Redelinghuys, 2013).

Conclusion

The water-related impacts of UOG extraction interlink across socio-economic and biophysical spheres. The challenges this presents necessitate an approach that recognizes the interrelated nature of impacts and that values ecological services to work with, rather than against, nature.

This article has looked at the reciprocal and multidimensional impacts of UOG extraction within the PED-nexus in South Africa, in terms of energy–water linkages, water–agriculture linkages and water–human population linkages. It is hoped that understanding these linkages will aid in the development of integrated policy for the protection of environmental resources and the human population, while promoting sustainable UOG extraction. We emphasize that UOG regulators need to address the spatial and temporal issues of scale in the institutional management of all the resources that interlink with water resources and that may be affected during UOG extraction.

In terms of water policy specifically, a few factors are important to ensure effective water resource protection and management. Institutions on different levels need to coordinate their management efforts to ensure that water resources are managed coherently across geographical and institutional boundaries during UOG extraction. Only multi-tiered management arrangements, with laws, policies and organizations that operate across administrative boundaries to manage water and other resources, can ensure proper protection of water resources. The potential of UOG extraction to have widespread and diffuse water-related impacts necessitates developing and implementing stringent regulations to protect water resources and perform proper monitoring of water resources before, during and after UOG extraction.

Proactive integrated management of the possible impacts of UOG extraction is especially relevant for South Africa, considering that it has yet to embark on this path.

References

Adams, M. B. (2011). Land and application of hydrofracturing fluids damages a deciduous forest stand in West Virginia. *Journal of Environment Quality*, *40*, 1340–1344. doi:10.2134/jeq2010.0504

Anderson, B. J., & Theodori, G. L. (2009). Local leaders' perceptions of energy development in the Barnett shale. *Southern Rural Sociology*, *24*(1), 113–129. Retrieved from http://www.landownerassociation.ca/rsrcs/SRS_2009_24-1-113-129.pdf

ANU (Australian National University). (2012). *Unconventional gas production and water resources: Lessons from the United States on better governance – A workshop for Australian government officials*. Canberra: Crawford School of Public Policy.

Barbee, J., Dutschke, M., & Smith, D. (2013, November 18). Botswana faces questions over licences for fracking companies in Kalahari. *The Guardian*. Retrieved from http://www.theguardian.com/environment/2013/nov/18/botswana-accusations-fracking-kalahari

Baynard, C. W. (2011). The landscape infrastructure footprint of oil development: Venezuela's heavy oil belt. *Ecological Indicators*, *11*, 789–810. doi:10.1016/j.ecolind.2010.10.005

Beemster, B., & Beemster, R. (2011). *Report on the effects of shale gas extraction by means of hydraulic fracturing in the Republic of Ireland* (2nd ed.). Sligo: Fracking Research and Information Centre.

Binnion, M. (2012). How the technical differences between shale gas and conventional gas projects lead to a new business model being required to be successful. *Marine and Petroleum Geology*, *31*(1), 3–7. doi:10.1016/j.marpetgeo.2011.12.003

Brasier, K. J., Filteau, M. R., McLaughlin, D. K., Jaquet, J., Stedman, R. C., Kelsey, T. W., & Goetz, S. J. (2011). Resident's perceptions of community and environmental impacts from development of natural gas in the Marcellus shale: A comparison of Pennsylvania and New York cases. *Journal of Rural Sociology*, *26*(1), 32–61. Retrieved from http://www.ag.auburn.edu/auxiliary/srsa/pages/Articles/JRSS%202011%2026%201%2032-61.pdf

Broderick, J., Anderson, K., Wood, R., Gilbert, P., Sharmina, M., Footitt, A., ... Nicholls, F. (2011). *Shale gas: An updated assessment of environmental and climate change impacts*. Tyndall Centre for Climate Change Research, University of Manchester. Retrieved from http://www.co-operative.coop/Corporate/Fracking/Shale%20gas%20update%20-%20full%20report.pdf

Broomfield, M. (2012). *Support to the identification of potential risks for the environment and human health arising from hydrocarbons operations involving hydraulic fracturing in Europe* (Report No. ED57281-17c). European Commission DG Environment. Retrieved from http://ec.europa.eu/environment/integration/energy/pdf/fracking%20study.pdf

Brynard, P. A., & Stone, A. B. (2004). From the Rio to Johannesburg world summits: On the road to policy implementation. In W. Fox & E. Van Rooyen (Eds.), *The quest for sustainable development* (pp. 22–45). Landsdowne: Juta.

Buschke, F. T., & Vanschoenwinkel, B. (2014). Mechanisms for the inclusion of cumulative impacts in conservation decision-making are sensitive to vulnerability and irreplaceability in a stochastically simulated landscape. *Journal for Nature Conservation, 22*(3), 265–271. doi:10.1016/j.jnc.2014.02.002

CER (Centre for Environmental Rights). (2013). *Minimum requirements for the regulation of environmental impacts of hydraulic fracturing in South Africa.* Cape Town. Retrieved from http://cer.org.za/wp-content/uploads/2013/12/CER-Minimum-Requirements-for-Regulation-of-Env-Impacts-of-Fracking-Dec-2013.pdf

Claassen, M. (2010). How much water do we have? In S. Oelofse & W. Strydom Eds., *A CSIR perspective on water in South Africa – 2010.* (CSIR Report No. CSIR/NRE/PW/IR/2011/0012/A. ISBN: 978-0-7988-5595-2). Pretoria, South Africa: Council for Scientific and Industrial Research.

Cohen, B., & Winkler, H. (2014). Greenhouse gas emissions from shale gas and coal for electricity generation in South Africa. *South African Journal of Science, 110*(3/4), 1–5. doi:10.1590/sajs.2014/20130194

Considine, T. M., Watson, R. W., & Considine, N. B. (2011). *The economic opportunities of shale energy development* (Energy Policy & the Environment Report, No. 9, June). The Manhattan Institute for Policy Research. Retrieved from http://www.manhattan-institute.org/html/eper_09.htm

Cooke, K. (2015, January 6). Plummeting oil price casts shadow over fracking's future. *The Guardian.* Retrieved from http://www.theguardian.com/environment/2015/jan/06/oil-price-casts-shadow-over-frackings-future

De Rijke, K. (2013). The Agri-Gas fields of Australia: Black soil, food and unconventional gas. *Culture, Agriculture, Food and Environment, 35*(1), 41–53. doi:10.1111/cuag.12004

De Wit, M. J. (2011). The great shale debate in the Karoo. *South African Journal of Science, 107* (7/8), 2–10. doi:10.4102/sajs.v107i7/8.791

Dolesh, R. J. (2011, June). Fractured parks. *Parks & Recreation,* pp. 56–61. Retrieved from http://ezine.parksandrecreation.org/HTML5/NRPA-Parks-Recreation-Magazine-June-2011

Dondeynaz, C., Moreno, C. M., & Lorente, J. J. C. (2012). Analysing inter-relationships among water, governance, human development variables in developing countries. *Hydrology and Earth System Sciences, 16*, 3791–3816. doi:10.5194/hess-16-3791-2012

Econometrix. (2012). *Karoo shale gas report: Special report on economic considerations surrounding potential shale gas resources in the southern Karoo of South Africa.* Econometrix (Pty) Ltd. Retrieved from http://us-cdn.creamermedia.co.za/assets/articles/attachments/38064_ksg_report_-_february_2012.pdf

EPRI (Electric Power Research Institute). (2002a). *Water and sustainability* (volume 3): *U.S. water consumption for power production – The next half century.* Technical report. (Report no 1006786). Palo Alto, CA: EPRI.

EPRI (Electric Power Research Institute). (2002b). *Water and sustainability* (volume 4): *U.S. electricity consumption for water supply and treatment – The next half century.* Technical report. (Report no 1006787). Palo Alto, CA: EPRI.

Esterhuyse, S., Kemp, M., & Redelinghuys, N. (2013). Assessing the existing knowledge base and opinions of decision makers on the regulation and monitoring of unconventional gas mining in South Africa. *Water International, 38*(6), 687–700. doi:10.1080/02508060.2013.818478

Grant, L., & Chrisholm, A. (2014). *Shale gas and water: An independent review of shale gas exploration and exploitation in the UK with a particular focus on the implications for the water environment*. The Chartered Institution of Water and Environmental Management (CIWEM). Retrieved from www.ciwem.org/shalegas

Groenewald, C. J. (2011). A system approach to training in population, environment and development – South Africa. In UNDP (Ed.), *Volume 19: Experiences in addressing population and reproductive health challenges. Sharing innovative experiences* (pp. 159–190). Retrieved from http://tcdc2.undp.org/GSSDAcademy/SIE/Docs/Vol19/SIE.v19_CH7.pdf

Harrison, D. (2010, January 24–26). *An overview of health and health care in South Africa 1994-2010: Priorities, progress and prospects for new gains*. Discussion document on Health in South Africa 1994-2010. Muldersdrift: National Leader's Retreat.

Hedden, S., & Cilliers, J. (2014). *Parched prospects: The emerging water crisis in South Africa* (African Futures Paper 11). Institute for Security Studies. Retrieved from http://www.issafrica. org/uploads/AF11_15Sep2014.pdf

Hightower, M., & Pierce, S. A. (2008). The energy challenge. *Nature, 452*(20), 285–286. doi:10.1038/452285a

Holahan, R., & Arnold, G. (2013). An institutional theory of hydraulic fracturing policy. *Ecological Economics, 94*, 127–134. doi:10.1016/j.ecolecon.2013.07.001

Hummel, D., Adamo, S., de Sherbinin, A., Murphy, L., Aggarwal, R., Zulu, L.,... Knight, K. (2013). Inter- and transdisciplinary approaches to population–environment research for sustainability aims: A review and appraisal. *Population and Environment, 34*(4), 481–509. doi:10.1007/s11111-012-0176-2

IEA (International Energy Agency). (2012). *Golden rules for a golden age of gas*. Paris: International Energy Agency. Retrieved from http://www.worldenergyoutlook.org/media/weo website/2012/goldenrules/weo2012_goldenrulesreport.pdf

Lechtenböhmer, S., Altmann, M., Capito, S., Matra, Z., Weindrorf, W., & Zittel, W. (2011). *Impacts of shale gas and shale oil extraction on the environment and on human health* (IP/A/ENVI/ST/2011-07, June). Brussels: European Parliament. Directorate General for Internal Policies. Policy Department A: Economic and Scientific Policy. Retrieved from http://www.europarl.europa.eu/ document/activities/cont/201107/20110715ATT24183/20110715ATT24183EN.pdf

Millar, J., & Roots, J. (2012). Changes in Australian agriculture and land use: Implications for future food security. *International Journal of Agricultural Sustainability, 10*(1), 25–39. doi:10.1080/14735903.2012.646731

Milton, S. (2012, October 14–17). *Potential environmental risks and impacts of mining in the Karoo*. Paper presented at the Karoo Development Conference, Beaufort West, South Africa.

National Research Council. (2012). *Induced seismicity potential in energy technologies*. Washington, DC: The National Research Council. Committee on Induced Seismicity Potential in Energy Technologies Committee on Earth Resources, Committee on Geological and Geotechnical Engineering, Committee on Seismology and Geodynamics Board on Earth Sciences and Resources, Division on Earth and Life Studies. The National Academies Press.

O'Riordan, T. (2007). Faces of the sustainability transition. In J. Pretty, A. S. Ball, T. Benton, J. S. Guivant, D. R. Lee, D. Orr,... H. Ward (Eds.), *The SAGE handbook of environment and society* (pp. 325–334). London: SAGE.

PASA (Petroleum Agency of South Africa). (2015). *Exploration maps*. Retrieved from http:// www.petroleumagencysa.com/index.php/maps

Pelser, A. J. (2012). Health and environment. In H. C. J. Van Rensburg (Ed.), *Health and health care in South Africa* (2nd ed., pp. 189–234). Pretoria: Van Schaik.

Pelser, A. J., & Redelinghuys, N. (2008). *Trends, changes and challenges at the interface of population, environment and development in South Africa*. Pretoria: Department of Social Development.

Rabe, B. G., & Borick, C. (2013). Conventional politics for unconventional drilling? Lessons from Pennsylvania's early move into fracking policy development. *Review of Policy Research, 30*(3), 321–340. doi:10.1111/ropr.12018

Redelinghuys, N. (2012). Health and health status of the South African population. In H. C. J. Van Rensburg (Ed.), *Health and health care in South Africa* (2nd ed., pp. 237–290). Pretoria: Van Schaik.

RSA (Republic of South Africa). (1996). *The constitution of the Republic of South Africa, Act 108 of 1996* (Government Gazette No. 25799, 2 December, 2003). Cape Town: Government Printer.

RSA (Republic of South Africa). (1998a). *National Environmental Management Act, Act 107 of 1998.* Pretoria: Government Printer.

RSA (Republic of South Africa). (1998b). *National Water Act, Act 36 of 1998* (Government Gazette No. 19182, 26 Augustus 1998). Cape Town: Government Printer.

RSA (Republic of South Africa). (2002). *Mineral and Petroleum Resources Development Act, Act 28 of 2002.* Cape Town: Government Printer.

RSA DSD (Republic of South Africa. Department of Social Development). (2010). *Progress review of the implementation of: The White Paper on Population Policy for South Africa (1998) and the ICPD Programme of Action (1994).* Pretoria: DSD.

RSA DWA (Republic of South Africa. Department of Water Affairs). (2009). Water for growth and development framework (Version 7). Pretoria: DWA. Retrieved from https://www.dwaf. gov.za/WFGD/documents/WFGD_Frameworkv7.pdf

RSA DWA (Republic of South Africa. Department of Water Affairs). (2013). *National water resource strategy 2: Water for an equitable and sustainable future.* Retrieved from https://www. dwa.gov.za/nwrs/LinkClick.aspx?fileticket=CIwWyptzLRk%3D&tabid=91&mid=496

RSA NPC (Republic of South Africa. National Planning Commission). (2013). *National development plan 2030: Our future – make it work.* Retrieved from https://www.environment.gov. za/sites/default/files/docs/national_development_plan_2030vision.pdf

SAIRR (South African Institute of Race Relations). (2014). *The South Africa Survey 2013/2014.* Pretoria: SAIRR.

Schafft, K. A., Borlu, Y., & Glenna, L. (2013). The relationship between Marcellus Shale gas development in Pennsylvania and local perceptions of risk and opportunity. *Rural Sociology, 78*(2), 143–166. doi:10.1111/ruso.2013.78.issue-2

Scheffer, M., Carpenter, S. R., Lenton, T. M., Bascompte, J., Brock, W., Dakos, V.,... Vandermeer, J. (2012). Anticipating critical transitions. *Science, 338,* 344–348. doi:10.1126/ science.1225244

Scott, C. A., & Pasqualetti, M. J. (2010). Energy and water resources scarcity: Critical infrastructure for growth and economic development in Arizona and Sonora. *Natural Resources Journal, 50*(3), 645–682. Retrieved from http://aquasec.org/pubs2010/Scott-Pasqualetti-2010-NexusAz-Son-NRJ.pdf

Scott, C. A., Pierce, S. A., Pasqualetti, M. J., Jones, A. L., Montz, B. E., & Hoover, J. H. (2011). Policy and institutional dimensions of the water-energy nexus. *Energy Policy, 39,* 6622–6630. doi:10.1016/j.enpol.2011.08.013

Siddiqi, A., & Anadon, L. D. (2011). The water energy nexus in Middle East and North Africa. *Energy Policy, 39*(8), 4529–4540. doi:10.1016/j.enpol.2011.04.023

Sovacool, B. K., & Sovacool, K. E. (2009). Identifying future electricity–water tradeoffs in the United States. *Energy Policy, 37*(7), 2763–2773. doi:10.1016/j.enpol.2009.03.012

Statistics South Africa. (2013). *General household survey 2013* (Statistical release P0318). Pretoria. Retrieved from http://www.statssa.gov.za/publications/P0318/P03182013.pdf

United Nations Department of Economic and Social Affairs. Population Division. (2007). *Rural population, development and the environment.* Retrieved from http://www.un.org/en/develop ment/desa/population/publications/pdf/urbanization/rural-2007.pdf

USA DOE (United States of America. Department of Energy). (2006). *Energy demands on water resources: Report to Congress on the interdependency of energy and water.* Washington, DC: USA Department of Energy.

USEPA (United States Environmental Protection Agency). (2011). *Coalbed methane extraction: Detailed study report* (Report No. EPA-820-R-10-022). Washington, DC: USEPA Office of Water. Retrieved from http://water.epa.gov/scitech/wastetech/guide/304m/upload/cbm_ report_2011.pdf

Vasak, S. (2008). *Inventory of transboundary aquifers and information flow*. Utrecht: International Groundwater Resources Assessment Centre (IGRAC). Retrieved from http://www.bgs.ac.uk/sadcreports/sadc2008vasakigracisarmtransboundaryaquifers.pdf

Vengosh, A., Jackson, R. B., Warner, N., Carrah, T. H., & Kondash, A. (2014). A critical review of the risks to water resources from unconventional shale gas development and hydraulic fracturing in the United States. *Environmental Science and Technology, 48*(15), 8334–8348. doi:10.1021/es405118y

Wait, R., & Rossouw, R. (2014). A comparative assessment of the economic benefits from shale gas extraction in the Karoo, South Africa. *Southern African Business Review, 18*(2), 1–34.

Walsh, B. (2011). The gas dilemma. *Time, 177*(14), 40–48. Retrieved from http://connection.ebscohost.com/c/articles/59877316/gas-dilemma

Warner, B., & Shapiro, J. (2013). Fractured, fragmented federalism: A study in fracking regulatory policy. *Publius: The Journal of Federalism, 43*(3), 474–496. doi:10.1093/publius/pjt014

Warren, C. H. (2013). Shale gas in South Africa: Towards an understanding of the security implications. *African Security Review, 22*, 67–73. doi:10.1080/10246029.2013.766443

Williams, J., Stubbs, T., & Milligan, A. (2012). An analysis of coal seam gas production and natural resource management in Australia. Report prepared for the Australian Council of Environmental Deans and Directors by John Williams Scientific Services. Canberra: ACEDD. Retrieved from http://aie.org.au/AIE/Documents/Oil_Gas_121114.pdf

Woodford, A., Rosewarne, P., & Girman, J. (n.d.). *How much groundwater does South Africa have?* Retrieved from http://www.srk.co.za/files/File/newsletters/groundwater/PDFs/1_A_Woodford.pdf

Zoback, M., Kitasei, S., & Copithorne, B. (2010). *Addressing the environmental risks from shale gas development* (Briefing paper 1). Washington, DC: Worldwatch Institute. Retrieved from https://www.worldwatch.org/files/pdf/Hydraulic%20Fracturing%20Paper.pdf

Lessons from Yanacocha: assessing mining impacts on hydrological systems and water distribution in the Cajamarca region, Peru

Diana Vela-Almeida, Froukje Kuijk, Guido Wyseure and Nicolas Kosoy

ABSTRACT

A major concern of mining activities is their influence on hydrological systems. This article highlights impacts on water flows and distribution in the Mashcon catchment in Cajamarca, Peru, one of those most affected by the Yanacocha mining project. Some important concerns are identified regarding changes in water flows, lowering of water tables, and decrease of base flows. These considerations indicate deficiencies in distributing actual water uses in relation to the allocation of water rights. Finally, the article discusses challenges for regulation of mining, including democratic processes for water management that require clear accountability in the context of local social needs.

Introduction

Water is arguably the most common receptor of environmental impacts, and the use of this rival resource creates a complex competition among multiple land uses (Bebbington, Bebbington, & Bury, 2010; Bebbington & Williams, 2008; Bridge, 2004). Mining activity is one such land use whose water demand impacts other uses in society. In regions where water is conceived as a valuable and scarce resource, conflicts emerge as a consequence of the disruption of traditional water uses. The Yanacocha gold mine, in the Cajamarca district of Peru, is a clear example of high levels of contestation related to multiple water uses. The Yanacocha mine has become one of the largest gold extraction projects in the world. The scale of its operations is defined by not only the amount of material extracted but also the associated volumes of water harnessed in the process. The project is located in the headwaters of the *jalca* highland ecosystem, which regulates water in the region by recharging and storing water in the soil and maintaining a supply downstream (Buytaert et al., 2006). Mining activities in this sensitive ecosystem have caused rapid alteration of hydrological regimes and several restrictions in the access to water for other legitimate users, resulting in multiple claims of water shortages (Rojas, 2010; Yacoub, 2013).

Multiple studies have analyzed the environmental impacts of the Yanacocha mine. These studies have focused on contested claims of water contamination (Yacoub, 2013), conflicts over water use, water management institutions, and accountability (Bebbington et al., 2010; Bebbington & Williams, 2008; Bury, 2004, 2005; Sosa & Zwarteveen, 2014). However, scant research has been devoted to understanding the impacts on the quantity of water resulting from competing water uses across the whole catchment. Although the identification of impacts is complex, estimation of the effect of mining on water flows represents a significant advance in the discussion of water access and distribution in Cajamarca. Thus, this article analyzes, first, how the use and manipulation of water by mining activities has potentially altered the quantity of water in the catchment; secondly, it addresses the implications for distribution among different water users. These inquiries are important for developing mechanisms of accountability to maintain water resources, regulate mine sites, and secure access to water that aligns with social preferences.

Competition over the limited water resource represents a substantial liability to users, especially in the dry season. The three main water users in the district of Cajamarca are small farmers, the urban population and the Yanacocha mining company. Small farmers are the largest group of water users in the region, representing 67% of the population (INEI, 2007). Traditionally, these rural communities depend heavily on water-intensive agriculture and livestock activities to survive (Garcia & Gomez, 2006). The second-largest group of water users is the growing urban population in the downstream city of Cajamarca, which depends primarily on water supply influenced by the Yanacocha project. Finally, the Yanacocha mine itself is considered the third-most influential user due to the sheer volume of water managed for operations. Accordingly, the distribution of water among several users requires an open discussion to define priorities based on social understandings of collective well-being, as the allocation of a rival resource might not necessarily satisfy all water users (Falkenmark, Lindh, de Mare, & Widstrand, 1980).

In the next section, the main quantitative alterations of water are described as well as managerial changes in the distribution of water occurring in Cajamarca over the past 20 years. The third and fourth sections present the case study and the methods used for analysis. The results presented in the fifth section indicate that alteration of the hydrological regime is occurring as a result of large-scale water mobilization by the mining company. The reduction of water flows is highlighted, which affects water supply for irrigation purposes as well as future hazards for domestic water supply. Finally, the complexity of physical processes associated with mining is discussed, which renders the determination of precise impacts on water flows a challenging yet non-inconsequential process, requiring increased scrutiny and democratic mechanisms for sound accountability.

Mining activities, water impacts and distribution

Physical alterations of water resources

Large-scale mining operations alter the hydrological regime by transforming the landscape and affecting the quality and quantity of water resources throughout the catchment (Bridge, 2004; Younger & Wolkersdorfer, 2004). The impacts are intensified as mineral reserves are often located in vulnerable glacial or headwater areas of drainage basins that serve as supply for downstream populations. In Peru, for example, historical records of observed environmental impacts have contributed to the reputation of large extractive activities as responsible for affecting water resources (Scurrah, 2008).

In mining activities, water is mainly used as an inflow for ore separation procedures, machinery washing and dust control, with gold mining using the most water per unit of material produced (Mudd, 2007). According to Mudd (2007), between 1991 and 2006, the global estimated average water flow for gold mining was 691 m^3 per kilogramme of gold. However, the total amount of water harnessed by a mining project is greater than the water consumed in these operations; mining companies also extract groundwater to discharge excess from the mine site. Open-pit extraction requires dewatering, a process in which groundwater is pumped out of the surrounding aquifers to lower the water table below the base-floor of the pit. Dewatering is done to reduce risk of landslide and acid drainage (Sperling, Freeze, Massmann, Smith, & James, 1992). This transfer of water further alters the hydrological regime of the catchment by increasing surface runoff and decreasing base flows in streams.

Mining activities also prevent aquifer recharge by altering the natural upstream drainage system. For instance, deviation canals and sinks are built around open pits to intercept precipitation and prevent the water table from rising (Sperling et al., 1992). Aquifers, however, play an important role by maintaining the hydrological regulative process of the watershed, which requires rainfall recharge, as well as sustaining the base flows of streams during the dry season (Todd & Mays, 2005). The alterations to the drainage system caused by mining present a risk of post-mining flooding, uncontrolled discharge of stored water, and drought (Younger & Wolkersdorfer, 2004).

Hydrological analysis thus permits the assessment of the state of the resource and depicts how critical dynamics of access and use are influenced by mining activities. As Kemp, Bond, Franks, and Cote (2010) discuss, apart from the work on technical frameworks for water use, quantification and accountability within the mining sector (see Cote, Moran, Cummings, & Ringwood, 2009, for more details), there should also be integration with studies of the implications of mining water use for water distribution and management.

Socio-political alterations of water access and distribution

Along with physical alterations, hydrological processes are constantly shaped by social changes (Budds, 2009). New development projects, extractive industries, and water policies modify previous forms of water resources management and subsequently the distribution of water among users (Bebbington, 2013; Budds & Hinojosa-Valencia, 2012). Specifically, the involvement of the private sector alters previous forms of water governance through the appropriation of water resources. Although the state exercises control over the allocation of water rights in Peru, these rights are re-allocated by illegitimate transactions between the private mining company and irrigation-water users, resulting in unaccountable environmental governance. Such actions are often supported by neoliberal extractive policies, accommodating water authorities and the exclusion of community demands (Achterhuis, Boelens, & Zwarteveen, 2010; Budds & Hinojosa-Valencia, 2012; Bury, 2005).

In Cajamarca, for instance, there have been cases in which irrigation-water users in the area of influence of Yanacocha have been dispossessed of their water rights through direct negotiations, bribery and economic compensation, without oversight or intervention by the state. In this process, irrigation water users renounce their water rights and transfer them to the mining company, with minimal accountability (Sosa & Zwarteveen, 2012, 2014). Thus, water rights are shifted from traditional users to the mining company, and the Yanacocha mining company assumes de facto control over the resources, thus inciting social tensions among local people (2012, 2014). Such re-allocations affect livelihoods and appropriate the means of subsistence of rural people (Bebbington, 2007; Bebbington, Hinojosa, Bebbington, Burneo, & Warnaars, 2008; Bury, 2004, 2005). Accordingly, water management requires accounting for the impacts on water resources, which in turn have profound implications for the creation of conflicts over the distribution of water between different users in the catchment (Bebbington & Williams, 2008). In the following section, the case study is contextualized in order to present findings on water use and distribution in Cajamarca.

Case study

The Yanacocha project in Cajamarca, Peru, is one of the largest gold mines in Latin America, with gold production of 498,000 troy ounces in 2014 (Newmont, n.d.). The mine concession encompasses 172,500 ha, and operations cover approximately 15,500 ha (MYSRL, 2006). This project started in 1993 under the administration of Minera Yanacocha (MYSRL), a consortium between Newmont Mining Corporations, Buenaventura, and the World Bank's International Finance Corporation. The Yanacocha complex consists of 13 open-pit mines, nine rock residue heaps and four lixiviation piles in six main zones: Carachugo and Maqui Maqui (already closed), San José and Cerro Negro (already mined), and Yanacocha and La Quinua–El Tapado (in operation).

The mine site is at the peak of the Yanacocha Mountain, in the north Andean highlands, at an elevation of nearly 4000 m. Operations are positioned on headwaters of four important catchments in this area: Chonta, Mashcon, Rejo, and Honda. For this study, the Mashcon catchment was chosen for analysis because mining operations are primarily located in the upstream portion of this catchment and it is the most densely populated in the area of influence of the mine (MYSRL, 2006, 2007). The Mashcon catchment is to the south and south-east of the Yanacocha mine, as illustrated in Figure 1. The catchment is approximately 15,820 ha in area, ranges from 2700 to 4100 m in altitude, and is formed of two sub-catchments, for the Grande and the Porcon Rivers, which join downstream to form the Mashcon River before reaching the city of Cajamarca. Around 6500 people directly depend on the catchment as its main rivers and streams provide water to small rural communities and to the majority of the population in the city of Cajamarca (Rojas, 2010).

Cajamarca is currently experiencing social unrest due to environmental impacts connected to MYSRL. Since 2000, the Observatory of Mining Conflicts in Peru (Observatory of Mining Conflicts in Peru, n.d.) has reported water-related conflicts associated with the destruction of irrigation canals, reduction of flows, and water contamination, which have resulted in multiple protests against MYSRL. For its part,

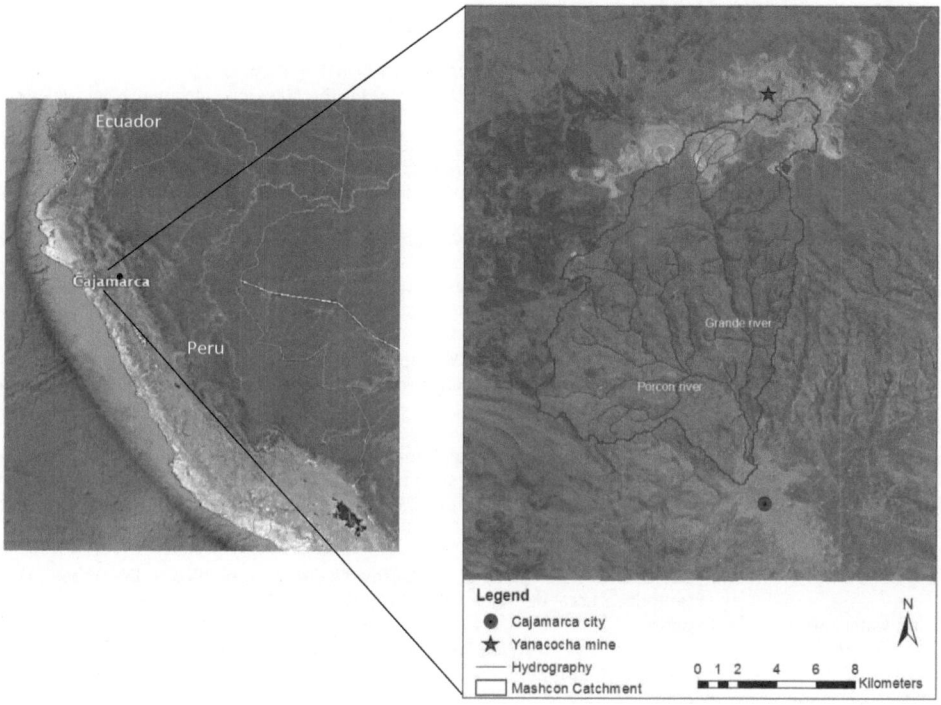

Figure 1. Location of the Mashcon catchment in the Cajamarca District, Peru.

the company argues that sophisticated technological processes to avoid water impacts are being employed to prevent mining activities from affecting other land practices (MYSRL, 2011a). MYSRL stresses that mitigation plans and monitoring ensure that the water supply is not being reduced downstream during the dry season. Conflicts are likely to be exacerbated due to growing population and demands for just access to resources, as well as dynamic environmental conditions such as climate change and the presence of recurrent droughts.

Methods

The analysis of hydrological changes and water distribution in the Mashcon catchment is based on secondary information provided by multiple organizations involved in water management, monitoring and planning in the Cajamarca district. Information was collected from the databases of local and regional authorities and water user associations, as well as from literature reviews of independent studies, environmental impacts assessments, and MYSRL reports. Table 1 details the sources of data used in this study. Information was processed to identify alterations of the hydrological regime due to mining activities, with definitions of different water uses and the distribution of water based upon use and established water rights. Descriptive statistics were conducted where possible to determine percentages of water flow changes in the river, proportion

Table 1. Sources of hydrological data collection, water rights and water used by different sectors.

Water flows of the Grande River	Reports: 'Optimized Master Plan SEDACAJ 2012–2017' (SEDACAJ, 2012) 'Inventory of sources of surface water for the Mashcon catchment' (Benavides et al., 2007) 'Evaluation of hydrological characteristics of the Porcon and Grande Rivers in the Province of Cajamarca. Hydrological year 2010–2011' (SENAMHI, 2011) 'Supplemental environmental impact assessment Yanacocha Oeste' (MYSRL, 2006) Raw data: River flow of the Grande River, provided by the Local Water Authority in Cajamarca
Information on streams and springs	Report: 'Inventory of sources of surface water for the Mashcón catchment' (Benavides et al., 2007) Raw data: Database discharges on MYSRL's discharge outlets and receptor water bodies, provided by the Group for Capacity Building and Intervention towards Sustainable Development, GRUFIDES
Groundwater information	Report: 'Supplemental environmental impact assessment Yanacocha Oeste' (MYSRL, 2006) 'Environmental impact assessment: exploration project Maqui Maqui' (MYSRL, 2010b)
Meteorological information	Precipitation values: database provided by the National Service of Meteorology and Hydrology of Peru, SENAMHI.
Water rights distribution	Raw data: Database of water rights of the Cajamarca district from 1985 to December 2013 (Autoridad Local del Agua Cajamarca [ALA-C], 2013)
Mining water use	Reports: MYSRL's sustainability reports from 2008 to 2013 (MYSRL, 2008, 2009, 2010a, 2011b, 2012, 2013, 2014)
Domestic water use	Reports: 'Optimized Master Plan SEDACAJ 2012-2017' (SEDACAJ, 2012). 'Inventory of sources of surface water for the Mashcón catchment' (Benavides et al., 2007)
Irrigation water use	Report: 'Supplemental environmental impact assessment Yanacocha Oeste' (MYSRL, 2006) Raw data: Database of irrigation canal network by the Association of Water Users of the Mashcon catchment Database of discharge water flows of irrigation canals by the Monitoring Commission of Irrigation Canals in Cajamarca, COMOCA

of streams and springs affected, lowering of water tables and distribution of water rights among users.

For the discharge of water in the monitored irrigation canals, a time series analysis was carried out to test temporal trends in water flows. The non-parametric Mann-Kendall test (Mann, 1945) was used since data exhibited a non-normal distribution. The Mann-Kendall test is a monotonic trend regression analysis that shows increasing or decreasing trends over time when there is a consistent change that is not necessarily linear. This test replaces a parametric linear regression analysis when the data are not normally distributed. The Mann-Kendall test assumes that all values are independent and that the distribution of values is constant and sufficiently large that no correlation exists across different time periods. The test was conducted using information for the dry season (June–October) for the past 14 years. Despite the dry season's extending from May to September in Cajamarca, for this analysis, a delay in measuring base flows more realistically assesses groundwater following a reduction in precipitation.

Finally, the biggest challenge in making clear estimations of impacts on hydrological regimes and water availability is access to information. Most hydrological data for the upstream catchment area are limited and managed exclusively by MYSRL. However, the available information provides guidance for valid estimates of the effects of mining in the catchment and legitimate concerns regarding the implication of these effects on the availability of water for other uses.

Quantitative changes in water resources in the Mashcon catchment

The headwaters of the Mashcon catchment are subject to heavy environmental impacts due to mining extraction (Benavides, Ángeles, Salazar, & Abásolo, 2007; MYSRL, 2006). The sections below detail how landscape transformations have severely altered the water tables and the flow of rivers, streams and springs in the Mashcon catchment.

Impacts on groundwater

Mining activities affect groundwater hydrology by dewatering the area below the pit for depressurization. Precipitation falling to the surface of the mine site is rerouted from its natural drainage system, preventing rainfall infiltration. Instead, water is captured, treated, stored in artificial reservoirs and eventually drained to other outlets, irrespective of the original water bodies. Impermeability of the mining area, which prevents aquifer recharge, and permanent dewatering are direct causes of the lowering of water tables. In turn, the lowering of water tables below the depth of the spring outlet or streambed prevents groundwater discharge and reduces river base flows (Younger & Wolkersdorfer, 2004). Although MYSRL permanently monitors the water table at different gauging points along its installations, information on the influence of dewatering on the water tables is kept confidential. There is no clear identification of water tables, though according to MYSRL (2006), they have suffered a permanent decrease. Most of the open pits reach a depth of around 300 m beneath the natural topography, and in some cases reach a maximum of 500 m (MYSRL, 2010b). Since the water table must be below the base level of the pits, water tables in the aquifers need to be lowered dramatically. Table 2 shows that the water tables of the La Quinua, Yanacocha and Carachugo projects have fallen more than 100 m in less than a decade. To the authors' knowledge, more recent data are not available.

MYSRL has developed a mitigation programme for the reduction of groundwater in the Mashcon catchment. This programme includes collecting water and discharging it at specific points in the drainage system to restore lost flow in streams and irrigation canals (MYSRL, 2006). Despite the objectives of the mitigation programme, groundwater is continuously pumped and transferred to the surface at a rate much higher than the natural discharge rate, a process that generates a loss of hydrological balance in the catchment boundaries. However, MYSRL's reports do not mention any measures to remediate the loss of groundwater and restore the infiltration capacity of the soil.

Impacts on the Grande River

The greatest mining impacts are occurring in the Grande River subcatchment of the Mashcon catchment, since most of the upstream area is being obstructed by the mine

Table 2. Lowering of water tables of aquifers.

Project	Start of operations	Closure of operations	Average initial water table height (m)	Average latest water table height (m)	Change in height (m)
Cerro Negro	2003	2005	3678 (2001)	3590.3 (2004)	−87.5
La Quinua	2001	2007	3798 (1999)	3664.4 (2005)	−133.57
El Tapado	2007	2010	3545	3460 (n.a.)	−85
Corimayo	2011	2014	3554 (2001)	3543.8 (2005)	−9.9
Yanacocha Norte	1997	2009	4024 (1996)	n.a.	−65 m between 2000 and 2004
Yanacocha Sur y Oeste	1997	2009	4006 (2001)	3905.95 (2005)	−100.31
Carachugo	1993	2002	3934 (1998)	3828.5 (2004)	−105
Chaquicocha[a]			4037 (2001)	3956.84 (2002)	−80.3

Note: [a]Years of starting and closure of operations are not known.
Source: Information was taken from available data provided in the Yanacocha West Environmental Impact Assessment and annexes (MYSRL, 2006).

installations (MYSRL, 2006). The available river flow data for the Grande River are shown in Table 3. Due to scarcity of data and multiple locations for measuring flow, conclusive results are not possible; however, some concerns can be identified. The two periods with the most flow data are 2002 and 2013–2014. A simple comparison between these two monitored events shows an increase in flow of 35–70% during the rainy season. This increase in flow might be a result of increased surface water runoff from higher dewatering rates and denuded surfaces. A time series regression analysis suggests that precipitation is not a factor in the increase in flow in the Grande River as precipitation values do not differ significantly between 2002 and 2013 ($p = 0.2486$). The increase in water flow might represent a change in the natural hydrological regime of the catchment, which renders the area more vulnerable to floods and represents a potential risk for uncontrolled discharge of stored groundwater.

Due to alterations in the drainage system that affected the supply of water in the Grande River during the dry season, MYSRL and the local water authority in Cajamarca (ALA-C) agreed in 2005 on a minimum flow discharge rule of 0.50 m^3/s from a reservoir managed by MYSRL (Kuijk, 2015). Table 3 indicates that the minimum discharge of 0.50 m^3/s agreed by Yanacocha was not met in June and October 2014, two critical months in the dry season when water is needed for irrigation and domestic

Table 3. Available monitored records of flow (m^3/s) in the Grande River.

Year	Jan.	Feb.	Mar.	Apr.	May	June	July	Aug.	Sept.	Oct.	Nov.	Dec.
2002	0.74	0.96	1.20	0.92	0.66	0.41	0.31	0.29	0.32	0.49	0.44	0.69
2007							1.23					
2010										0.67	0.87	0.98
2011	2.10	2.41	1.24	1.49								
2013					2.63	1.87	1.05	0.75	0.79	0.62	0.70	0.63	1.42
2014	1.55	2.14	2.88	1.30	1.30	0.46				0.29		
2014*			1.63	1.05				0.89				
2015		2.71										

* The two rows for 2014 represent the measurements of water flow from two different organizations: the first from ALA-C, the second from SENAMHI.
Source: Information was obtained from Autoridad Local del Agua Cajamarca – ALA-C (2013); SEDACAJ (2012); Servicio Nacional de Meteorología e Hidrología del Perú – SENAMHI (2011); and Benavides et al. (2007). The table presents only the data that have been made publicly available.

consumption. It is necessary to emphasize however that the lack of data for other years precludes confidence in compliance with the agreement on minimum discharge.

Impacts on streams and springs

A detailed inventory of the streams contributing to the Porcon and Grande Rivers was conducted by Benavides et al. (2007). Figure 2 is a map of the main streams in the Mashcon catchment based on the inventory. The two most impacted streams are the Encajon and the Callejon, which in turn are the main tributaries of the Grande River

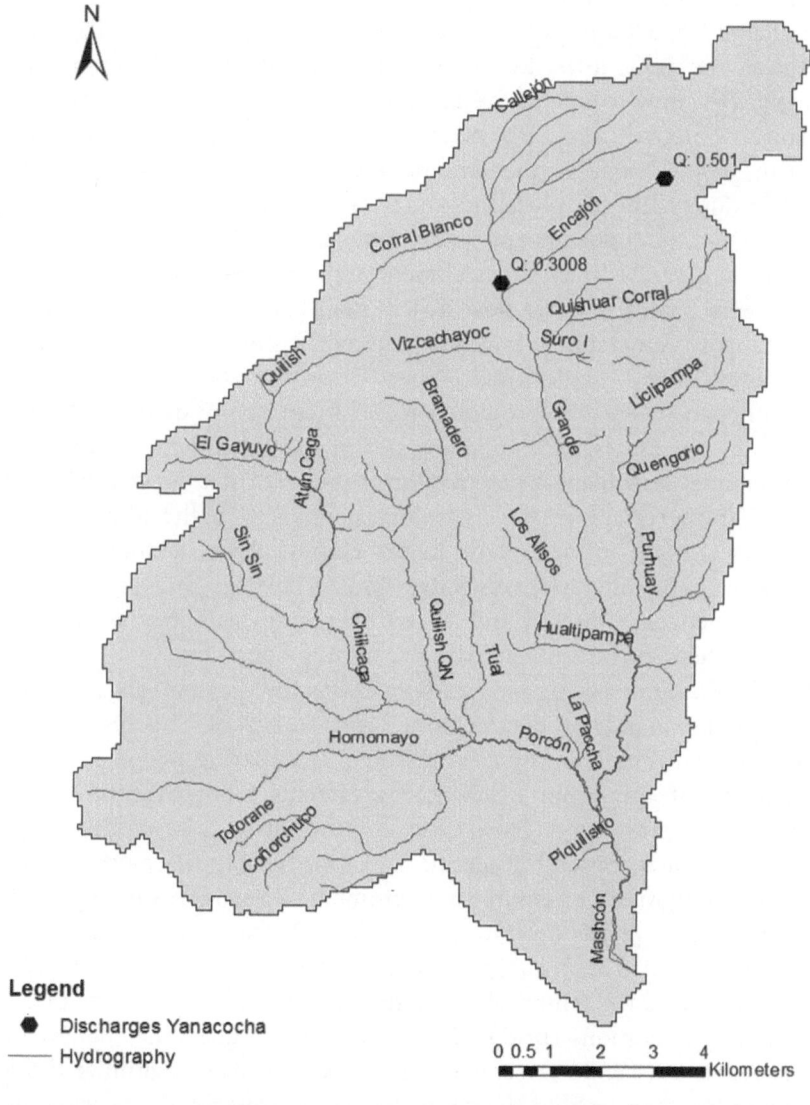

Figure 2. Hydrography of the Mashcon catchment identifying main streams. Additionally, two discharge points indicate the volume of water discharged (m^3/s) to the Callejon and Encajon streams by MYSRL.

upstream. The Encajon stream contributes to the Grande River with 60% of the flow in the dry season, while information on the contribution of the Callejon stream is not available (2006). Both streams have been disturbed by mining activities since 1994 as a result of operations in their drainage area. As a compensatory measure for the impact on natural drainage patterns of the streams, MYSRL agreed to discharge water from a reservoir to the Encajon and Callejon streams. Information provided by the Group for Capacity Building and Intervention towards Sustainable Development (GRUFIDES) on MYSRL's outlets and receptor water bodies between 2010 and 2012 indicates that the Encajon and Callejon streams receive a much higher discharge from the mine than the mean average yearly flow. For example, the mean average yearly flow of the Encajon stream is 0.0186 m^3/s, while the MYSRL's average discharge is 0.5013 m^3/s. Similarly, the Callejon stream has a mean average yearly flow of 0.127 m^3/s, while the MYSRL's average discharge is 0.3008 m^3/s. Although these flows represent average yearly values and will logically be much lower in the dry season, the increase of volume in the main streams changes the flow regime of the catchment, which is otherwise highly dependent on seasonality. Moreover, only the main streams that receive the discharge of the operation can supply water to the Grande River downstream, while small streams tend to dry up due to the overall falling water tables.

Benavides et al. (2007) also inventoried springs in the Mashcon catchment. Most of the springs in the Grande River subcatchment are in the mid-to-upstream catchment, with 93% of them having a water flow of less than 0.001 m^3/s. These two conditions make springs highly vulnerable to lowered water tables since they will no longer receive supply from aquifers.

In sum, although the results presented here cannot firmly quantify the impacts of mining on water nor assess the available water in the catchment, disturbances in the hydrological regime are evident. There are thus some critical interpretations that can be inferred here. These results on the hydrological impacts in the Mashcon catchment indicate sudden changes in water flow in the Grande River, a dramatic lowering of water table levels, and alteration of water flow in springs and streams. Moreover, upstream water networks directly influenced by mining activities are at the greatest risk of damage due to direct influence by the mining operation.

In 2000, MYSRL more strictly defined regulations for water management as a result of increasing local demands on the declining water resources. This resulted in a public demand for an environmental impact assessment prioritizing regional hydrology and the establishment of minimum discharge agreements from MYSRL (Kuijk, 2015). Nevertheless, such assessments have been insufficient to assess the complexity of mining effects on water. Several academic and research reports have shown that the hydrogeological component of environmental impact assessments tends to be generalized and minimizes emphasis on impacts on water (Moran, 2003; Rojas, 2010). These assessments lack sufficient information and analytical depth, which results in imprecise evaluations to diagnose the temporal and long-term hydrological effects of cumulative impacts of mining operations (Bridge, 2004). Li (2009) argues that the environmental impact assessments of MYSRL are conceived as mechanisms to legitimize the actions of the company rather than to hold the company accountable for the possible impacts of the operation. This is because the company itself defines the content of the document and identifies risks that are technically manageable. A critical liability is that each project in Yanacocha has its own environmental impact assessment that focuses on specific impacts related to individual projects to mitigate or compensate but does not account for the cumulative impacts on water, making it impossible to identify reinforcing feedback loops and synergies between different processes.

Finally, it should be highlighted that the MYSRL impacts on water are related not only to the reduction of water flow but also to the reallocation of water resources and disruption of the baseline hydrology. For instance, the soil removal from open pits where headwaters are located, alteration of the draining system, dewatering, and the development of reservoirs create a new hydrological regime shaped to meet the priorities of MYSRL. These impacts are the results of permissive regulations, nonexistence of public information, and permissive water authorities that have not been able to render MYSRL accountable.

Water distribution and users in the Mashcon catchment

Disturbances in water supply not only reflect the alterations in the hydrological regime but also highlight changes in water distribution. Water consumption in the Cajamarca district is divided among small farmers, the urban population, and MYSRL.[1] However, water is unequally distributed among these users. The distribution of water rights in the Mashcon catchment is presented in Table 4. The water rights granted to the three main users are an indicator of current imbalances in the distribution of water according to the social priorities for water supply.

The Peruvian Water Law establishes that the use of water must be prioritized for the satisfaction of social and environmental functions. However, precise operationalization of water distribution according to defined demands and availability remains uncertain. In practice, water is diverted to prioritize economic interests for mining; the Yanacocha mine uses 34.2% of the water demand in the Mashcon catchment (Table 4). This includes water consumed by the operations but also water transferred from aquifers to surface runoff. According to the licence for use of groundwater, MYSRL is authorized to pump 0.574 m^3/s of groundwater and use 0.060 m^3/s of surface water, for a total of 0.634 m^3/s. This volume extracted by MYSRL is more than 1.5 times the volume of water granted annually to the Cajamarca drinking water plant, which supplies water for 70% of the population of the city of Cajamarca (SEDACAJ, 2012). Although the dominant water use has been irrigation, representing 41.2% of the demand, the volume

Table 4. Multisectoral water rights based on licenses granted by the Autoridad Local del Agua Cajamarca in the Mashcon catchment. Estimates of the number of users and the area of irrigation and mining use were based on records of water rights until December 2013.

Use	No. of users	Area (ha)	Granted flow (m^3/s)	Percentage of demand
Irrigation use	7,869	1,669	0.7656	41.2
Domestic use	89,296*	n/a	0.4558**	24.6
Mining use	1	16,000	0.6344	34.2

* Based on the records of water rights granted upstream and estimated beneficiaries in the city of Cajamarca according to the estimated population in 2010 (127,363).

**The water licence of domestic use in the city of Cajamarca constitutes 0.200 m^3/s. Other values represent water rights in the catchment for multiple uses, of which domestic consumption is the main use.

of water granted for mining operations is much higher when considering a per user analysis since the water rights granted for irrigation are only 9.729×10^{-5} m^3/s per user.

MYSRL and its water use

MYSRL has been one of the most influential water users in Cajamarca. Given the economic priority that the Peruvian government has afforded to the mineral industry, MYSRL has been able to avoid critical accountability and closure of operations stemming from the socio-environmental repercussions occurring in the area. However, as specified, MYSRL has generated legitimate concerns for other water users. These concerns relate to the reduction of groundwater, scarcity of water in springs and streams in upstream areas, uncontrolled discharge of water by the Yanacocha mine to the Grande River and destruction of the natural hydrological regime.

In terms of the water used by MYRSL, the mine requires a constant supply of water. Table 5 shows the annual amounts of water managed in the mining installations since data was reported in 2008. The volume of total water consumption for all years is higher than the total water harnessed, indicating that the volume of water recycled in the installations is much larger than the yearly water inputs reflected in net water consumption. MYSRL argues that the company consumes approximately 2 million cubic metres (MCM), representing only 1% of the total water consumed in the region, in contrast to the amount of water used for agriculture, which is approximately 68 MCM (MYSRL, 2011a). However, MYRSL's total water consumption (Table 5) suggests a highly water-intensive process, which in most years uses almost double the quantity consumed by agriculture. Moreover, there is a progressive increase of total water harnessed, even though the last two years reported showed a reduction of the net water consumption.

Nearly all the water harnessed comes from groundwater. While the granting of water rights allows the company to pump 18.10 MCM per year (0.574 m^3/s) (Autoridad Local del Agua Cajamarca – ALA-C, 2013), the amount of water drained from aquifers since 2010 is greater than this. In the 2013 report, groundwater drainage surpasses the

Table 5. Reported water use in the Yanacocha mine, obtained from company sustainability reports (MYSRL, 2008, 2009, 2010a, 2011b, 2012, 2013, 2014). All figures in million cubic metres.

Year	Total water harnessed*	Surface drainage	Groundwater drainage	Precipitation	Total water consumption**	Net water consumption***	Water discharged
2008	29.69	7.90	11	10.79	n/a	0.36	29.34
2009	33.79	7. 95	13.47	12.36	125.10	5.60	31.44
2010	32.29	4. 90	18.29	9.10	126.97	2.14	34.48
2011	40.77	4. 14	27.05	9.58	128.64	n/a	46.37
2012	49.07	4. 22	32.40	12.45	89.92	4.12	44.96
2013	53.07	15.50	37.58	Included in surface drainage	62.27	3.30	49.78
2014	44.57	15.33	29.24	Included in surface drainage	47.82	2.67	41.89

* Comprises volumes from surface drainage, groundwater drainage, and precipitation.
** Includes volumes for recycled water used in the mining installations.
*** Values for 2008, 2012, 2013 and 2014 were calculated as the difference between the total water harnessed and the water discharged. For 2009 and 2010, net consumption was reported. Net consumption for 2011 was not calculated, as the value for water discharged was larger than the value for water harnessed.

company s water rights by more than twice the volume granted. These findings suggest that MYSRL greatly exceeds the permitted rights of water extraction, which might jeopardize access to water for other users that depend on flows upstream in the Mashcon catchment.

Domestic use for rural and urban populations

In 2012, only 68% of people in Cajamarca had access to permanent clean water, a problem mainly affecting rural populations (Instituto Nacional de Estadística e Informática – INEI, 2012). Around 9330 families live in the influence area of the mine and use water from streams and springs for domestic purposes (Sosa & Zwarteveen, 2012). However, the quality of water consumed by the rural population has been contested due to several claims of acid drainage and chemical discharges from the Yanacocha mine to the streams (Observatory of Mining Conflicts in Peru, n.d.). Descriptive statistics for the Mashcon catchment (Benavides et al., 2007) suggest that springs with a flow of less than 0.001 m^3/s are estimated to serve just one household; however, 204 springs with less than the previously stated volume provide water for more than one family. For example, the Chicos-Chicos spring has a hydrological yield of 0.0016 m^3/s and provides water to 103 families for domestic and agricultural use. Considering that 92% of the springs in the Mashcon catchment are traditionally used without official authorization and have not been registered, this might represent a problem for rural communities, as the loss of water flow in springs cannot be formally reclaimed. Although water supply is theoretically allocated according to defined water rights, this does not necessarily reflect actual demand. For example, upstream water for rural populations was neither documented nor clearly regulated, while water flows remain affected, possibly jeopardizing the livelihoods of rural populations of the Cajamarca district.

Due to the proximity of the mining operations to the city of Cajamarca, impacts also extend to the urban population. The water downstream is mainly directed to the city of Cajamarca and its 150,197 inhabitants (INEI, 2007). This population has been growing by 4%, a higher rate than in Peru's capital, Lima. The introduction of the Yanacocha project in 1993 is considered a major driver of rapid growth in the city, especially through the influx of new workers for the mine (Steel, 2013). As a consequence of this growth, the city has not been able to accommodate demands for water access. In addition, water shortages occurred regularly in 2011 and 2012, resulting in the MYSRL being accused of desiccation of the Grande River (Prado, 2012).

The population of the city of Cajamarca uses water provided by SEDACAJ, a company responsible for managing drinking water in the city. The SEDACAJ (2012) report identifies the Grande River as the main source of water for SEDACAJ, providing a supply representing approximately 75% of the river minimum flow (2012).[2] Since 1999, SEDACAJ's licence has been for 0.380 m^3/s; yet, domestic demand has increased due to population growth. In 2012, the SEDACAJ treatment plant had a supply flow of 0.250 m^3/s. Demand however, corresponded to 0.266 m^3/s, generating a deficit of 0.016 m^3/s. The water demand for the city of Cajamarca is predicted to grow from 0.266 m^3/s in 2011 to 0.646 m^3/s in 2035 (2012), a volume difficult to supply with current water sources. In fact, there is a paucity of alternatives for the expansion of supply sources for drinking water. Alternative sources for the future water supply for the city of Cajamarca are invariably influenced by MYSRL (SEDACAJ, 2012). Moreover, the mitigation plan of MYSRL for using discharge water from the mine would result in increased treatment costs for drinking water, a cost that will ultimately be shifted to the population of Cajamarca.

Small farmers' claims and water use

Small farmers in the region have traditionally been milk producers and cultivate small pastures and dairy cattle as their main agricultural activities. The historical use of surface water in the region is irrigation deriving from regulated canals in the catchment, which are managed by committees of water users. The distribution of water between different water associations is embedded in collective institutions for regulating the use and operations and maintaining the canal infrastructure. After persistent complaints from the upstream user committees regarding the reduction of their water flows in the canals, it was clear that MYSRL incited substantial competition over water.

Irrigation committees upstream of the Grande River depend on water diverted from most springs and streams. During the expansion of the La Quinua and Yanacocha mining projects, five canals in the subcatchment of the Grande River were impacted: the Encajon-Collotan, Yanacocha-Llagamarca, La Shacsha, Quishuar, and San Martín–Tupac Amaru–Río Colorado. The mining company agreed to supply 0.163 m^3/s of water from a mining reservoir to the affected canals to mitigate the effects of reduced base flows (MYSRL, 2006). However, this value does not represent a constant annual supply. In fact, water to the Quishuar and Encajon-Collotan canals was supplied for the 8.5 months of reduced precipitation (2006). Figure 3 shows the main canals in the Mashcon catchment together with the four discharge points for the Encajon-Collotan (0.042 m^3/s), Quishuar (0.056 m^3/s), Yanacocha-Llagamarca (0.025 m^3/s) and San Martín–Tupac Amaru–Río Colorado (0.040 m^3/s) canals (2006).

In return for the agreement made with MYSRL for a minimum water discharge in the canals, the irrigation committees gave up their water rights and accepted the company's conditions for supervision, provision and distribution of the water from the reservoir (Kuijk, 2015). These negotiations between water committees and MYSRL were carried out without involvement of the water authority (Sosa & Zwarteveen, 2012).

In order to ensure constant monitoring of the canals that could be affected by mining in the Mashcon catchment, a commission for the monitoring of irrigation canals (COMOCA) was created. Using COMOCA's monthly data on water flows of the canals, an historical trend analysis was conducted by using the Mann-Kendall test on the water flow supply. Results are summarized in Table 6. An increasing trend can be seen in the major upstream canals of Encajon Collotan, La Shacsha, and Quishuar, each located in the mining proximities of the Grande River subcatchment (see Figure 3). These upward trends could be the result of large volumes of water discharged from the mining operations as established in the mitigation agreement between MYSRL and the respective water associations. In contrast, four canals (Atunmayo, Carhuaquero

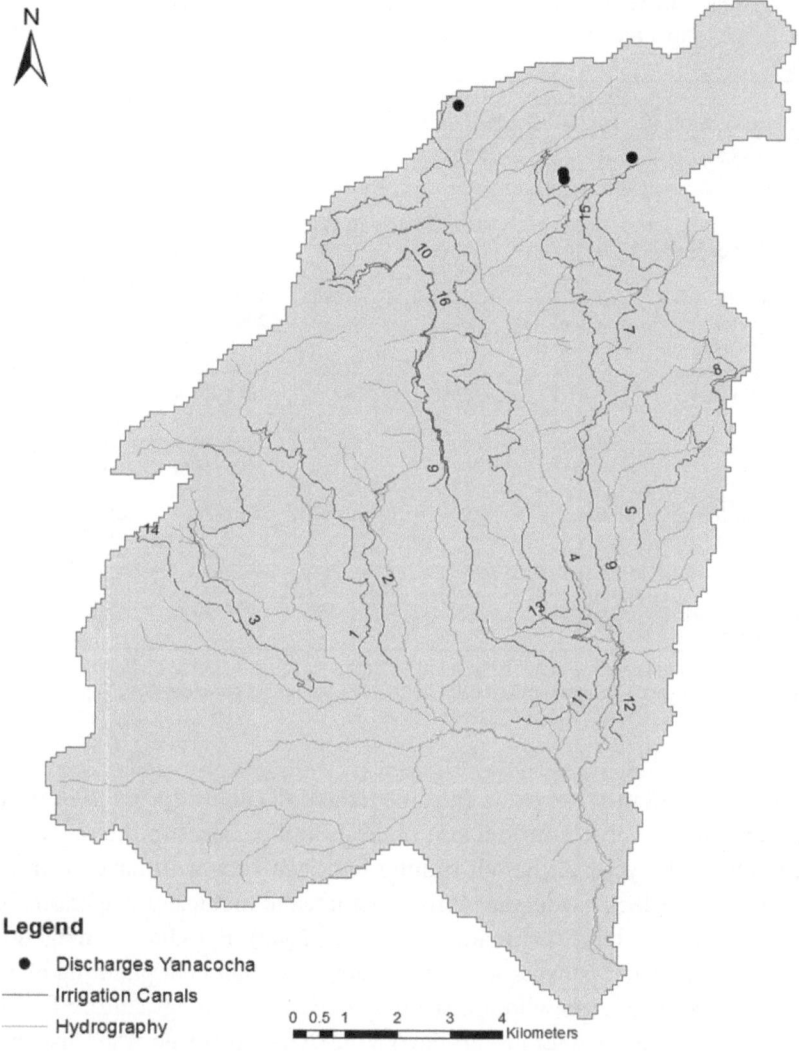

Figure 3. Primary irrigation canals in the Mashcon catchment. The canal numbers correspond to the reference numbers in Table 6.

Yacushilla, San Martín–Tupac Amaru–Río Colorado and San Salvador José de Coremayo) have a decreasing flow trend over the time period analyzed. Since each irrigation canal initially received water from streams, it is possible to deduce that the mining operations have affected the latter canals, which are dependent on water flows deriving from a series of networks of major upstream irrigation canals. Finally, the Cince Las Vizcachas, La Collpa and Quillish Porcon Bajo canals, in the less impacted Porcon River subcatchment, do not show a significant trend, suggesting that flow in those canals has not been affected. No information was available to analyze the flow trend for one of the main canals, the Yanacocha-Llagamarca.

Table 6. Irrigation canals within the influence area of the Yanacocha project in the Mashcon catchment. Results show a trend analysis on discharge water flows of the irrigation canals monitored by COMOCA.

Ref. no.	Irrigation canal	Water use right (m³/s)	Minimum flow (m³/s)	Maximum flow (m³/s)	Time series	Measuring point	p-value from trend analysis
1	Hermanos Cueva	0.0173	0.0012	0.12	2001–2014	CHCD-1	(–) 0.013*
						CHCI-1	(+) 0.00029***
						CHC	(–) 0.223
2	La Collpa	0.017	0.02	0.098	2001–2014	CCOL-1	(–) 0.395
3	Quilish Porcon Bajo	0.010	0.0015	0.151	2001–2014	CQUI-1	(+) 0.275
						CQUI-2	(–) 0.630
4	Atunmayo	0.020	0.0025	0.088	2003–2014	CAM-1	(–) 0.034*
5	Carhuaquero Yacushilla	0.0116	0.00058	0.150	2001–2014	CCY-1	(–) 0.00017***
6	Cince las Vizcachas	0.0014	0.0012	0.045	2004–2014	CCV-1	(–) 0.438
7	Encajón Collotan	0.0633	0.0068	0.071	2001–2014	CEC-1	(+) 2.22e-16***
						CEC-2	(+) 2.22e-16***
8	La Sacsha	0.00923	0.004	0.222	2001–2014	CSH-1	(+) 1.668e-06***
9	Quishuar	0.0848	0.003	0.150	2001–2014	CQ-1	(+) 1.192e-05 ***
						CQ-2	(+) 0.00496*
10	San Martin-Tupac Amaru-Río Colorado	0.163	0.001	0.563	2001–2014	CTU-1	(–) 1.500e-06***
						CTU-2	(–) 0.053
						CTU-2A	(–) 0.155
						CTU3	(–) 0.230
11	San Salvador José de Coremayo	0.068	0.0005	0.089	2004–2014	CSC-1	(–) 0.00045***

Note. Two-sided p-values indicated with *** for $p < .001$, ** for $p < .01$, and * for $p < .05$. Increasing and decreasing trends indicated by (+) and (–). Data from COMOCA 2001–2014, Autoridad Local del Agua Cajamarca – ALA-C (2013).

A decreasing trend in water flows in some irrigation canals during the dry season is of critical concern since optimal production of crops and pastures in the region depends on water throughout the year. Although mining has been shown to have an impact on the reduction of flow, the lack of adequate infrastructure, and inefficient irrigation methods, are also to blame (Bernet, Hervé, Lehmann, & Walker, 2002). For efficient irrigation technology to be implemented, it is necessary that greater investments in planning be made. Such burdens fall on poor farmers who are strongly dependent on agriculture for subsistence. Although current water policies in Peru favour so-called advanced water use technologies, more efficient infrastructure and stronger controls among water user associations to mitigate water reduction, these arguments divert the discussion from accountability for water impacts and shift liability to irrigation-water users.

Lessons learnt

Despite the critical importance of hydrological analysis to inform decision making, this field is hindered by an acute scarcity of data, which jeopardizes formulation of adaptive measures for water management (Buytaert et al., 2014). A critical factor that limits drawing conclusive results on the impacts on water of the Yanacocha mine is the lack of a baseline study of the region's natural hydrology, historical records of water flows and changes in the water tables of aquifers in the Mashcon catchment. This challenge has been confirmed by other authors (Kuijk, 2015; Yacoub & Cortina, 2007). However, the urgency of implementing regulations on the use and distribution of water also provides the opportunity to generate citizen mechanisms to define restrictions on the use of water in intensive land-transformation activities (2014).

The Yanacocha project will soon begin its overall closure stage; however, the post-mining phase is equally critical to secure water provisioning for wider demands from the catchment. As a result of the constant threat of water reduction, proposed policies are being crafted in favour of the efficient use of water (Boelens & Vos, 2012). These alternatives focus mainly on technological solutions such as the construction of reservoirs and dams that will supply water to different users (MYSRL, 2013). Construction of reservoirs and dams aims to mitigate the reduction of base flows but does not aim to regenerate water flows, thereby threatening the sustainability of the hydrological regime in the future (Fonseca, McAllister, & Fitzpatrick, 2013).

Management of water resources in Yanacocha is associated not only with physical changes, but also with poor management systems, the granting of water rights without adequate planning, and the lack of monitoring mechanisms to control actual use. Thus, changes in water distribution require more than new mechanisms for coping with the reduction of water; a thorough analysis of water sources, clear accountability of water use by multiple stakeholders, socially just distribution of water rights, and assumed responsibilities for water loss are also required.

The regulatory role of state institutions also requires further strengthening. Although the role of the state in managing water resources is essential, overly top-down water policies have a history of disciplining users and undermining diverse local systems of water management in several Andean countries (Boelens, 2009; Gelles & Boelens, 2003). Instead, recognition of 'legal pluralism' can serve to build stronger mechanisms of management, control and accountability for water resources. A broader regulatory context based on strengthening customary water rights should recognize diverse institutionally contingent arrangements (Vos, Boelens, & Bustamante, 2006). Moreover, the interplay of plural institutional mechanisms depends critically on deliberative processes allowing democratic decision making and incorporating the discussion of just distribution of water according to water availability and social priorities. A balanced decision-making system between technical analyses and recognition of a just and equitable system for allocation of water should set appropriate limits on water use and form the benchmark for monitoring compliance (Perreault, 2014; Zwarteveen & Boelens, 2014). There is thus an urgent need to *collectively* define, based on environmental criteria and population requirements, whether mining activities are appropriate and where they should not be permitted.

The Yanacocha project is reaching its final years of production, and the company is currently targeting a new project known as Conga. This project is an extension of MYSRL, 25 km north-east of the current mining district. Given a similar climate and ecological conditions, as well as historical records in terms of water management, it is necessary to develop a sound baseline in the catchments around the planned Conga project. Lastly, ethical questions should also be deliberated upon regarding water rights allocation and priorities for environmental conservation and livelihoods in the region of Cajamarca.

Conclusions

The cause-and-effect relationship between mining activities, impacts on water, and subsequent land uses is not straightforward and suggests the need for a more comprehensive understanding of the specific biophysical conditions and land use practices that affect the provisioning of water. The lack of baseline information and monitoring represents a challenge for the definition of a hydrological budget between water availability and demand in the Mashcon catchment. However, this lack of complete information regarding the modifications of water flows should not be a reason for the failure to identify salient and proactive policy solutions associated with mining extraction. We have argued that water reduction, and the lack of regulation in the use and distribution of water rights, are serious barriers to water sustainability in Cajamarca. Regulatory mechanisms to cope with water scarcity should be based on the recognition that water resource allocation is conditioned by both availability and a contestation of social priorities. The policy relevance lies in whether it is sustainable to assign water rights without hydrological planning and clear knowledge of the availability of water resources. A more comprehensive and socially just distribution of water requires critical modifications to ensure that local and regional development is not restricted to the success of mining activities, but can be a reality for the totality of the population in Cajamarca.

Notes

1. Additional water users relate to energy production and other industrial uses; however, they are not considered here, as they do not represent a significant value in terms of volume of water used or number of users.
2. Other small sources of water for SEDACAJ come from the Ronquillo River, which provides an additional flow of 0.060 m^3/s. In the peripheral zone of the city water is provided by small aquifers.

Acknowledgements

We are grateful to GRUFIDES, the Ecological and Economic Zoning (ZEE) project coordinators from the Gobierno Regional Cajamarca, Autoridad Local del Agua- Cajamarca, Junta de Usuarios de Riego de la Cuenca del Rio Mashcon, COMOCA, and Prof. Francisco Huamán of the Universidad Nacional de Cajamarca for valuable advice. We also thank Vijay Kolinjivadi, Kate Reilly, Carlos Cerdán and anonymous reviewers for their constructive review of previous drafts.

Funding

The authors acknowledge financial support from the National Secretary of Higher Education, Science, Technology, and Innovation (SENESCYT) of Ecuador.

References

Achterhuis, H., Boelens, R., & Zwarteveen, M. (2010). Water property relations and the modern policy regimes: Neoliberal utopia and the disempowerment of collective action. In R. Boelens, D. Getches, & A. Guevara-Gil (Eds.), *Out of the mainstream water rights, politics and identity* (pp. 27–50). London: Earthscan.

Autoridad Local del Agua Cajamarca - ALA-C. (2013). *Registro administrativo de derechos de uso de agua en el Distrito de Cajamarca*. Cajamarca: Autoridad Local del Agua Cajamarca.

Bebbington, A. (2007). *Minería, movimientos sociales y respuestas campesinas: Una ecología política de transformaciones territoriales* (Vol. 2). Lima: Instituto de Estudios Peruanos.

Bebbington, A. (2013). *Industrias extractivas: Conflicto social y dinámicas institucionales en la Región Andina*. Lima: Instituto de Estudios Peruanos.

Bebbington, A., Bebbington, D. H., & Bury, J. (2010). Federating and defending: Water, territory and extraction in the Andes. In R. Boelens, D. Getches, & A. Guevara-Gil (Eds.), *Out of the mainstream water rights, politics and identity* (pp. 307–329). London: Earthscan.

Bebbington, A., Hinojosa, L., Bebbington, D. H., Burneo, M. L., & Warnaars, X. (2008). Contention and ambiguity: Mining and the possibilities of development. *Development and Change, 39*, 887–914. doi:10.1111/j.1467-7660.2008.00517.x

Bebbington, A., & Williams, M. (2008). Water and mining conflicts in Peru. *Mountain Research and Development, 28*, 190–195. doi:10.1659/mrd.1039

Benavides, I., Ángeles, I. R., Salazar, E., & Abásolo, J. (2007). *Inventario de fuentes de agua superficial de la Cuenca del Mashcón*. Cajamarca: INRENA.

Bernet, T., Hervé, D., Lehmann, B., & Walker, T. (2002). Improving land use by slope farmers in the Andes: An economic assessment of small-scale sprinkler irrigation for milk production. *Mountain Research and Development, 22*, 375–382. doi:10.1659/0276-4741(2002)022[0375:ILUBSF]2.0.CO;2

Boelens, R. (2009). The politics of disciplining water rights. *Development and Change, 40*, 307–331. doi:10.1111/dech.2009.40.issue-2

Boelens, R., & Vos, J. (2012). The danger of naturalizing water policy concepts: Water productivity and efficiency discourses from field irrigation to virtual water trade. *Agricultural Water Management, 108*, 16–26. doi:10.1016/j.agwat.2011.06.013

Bridge, G. (2004). Contested terrain: Mining and the environment. *Annual Review of Environment and Resources, 29*, 205–259. doi:10.1146/annurev.energy.28.011503.163434

Budds, J. (2009). Contested H_2O: Science, policy and politics in water resources management in Chile. *Geoforum, 40*, 418–430. doi:10.1016/j.geoforum.2008.12.008

Budds, J., & Hinojosa-Valencia, L. (2012). Restructuring and rescaling water governance in mining contexts: The co-production of waterscapes in Peru. *Water Alternatives, 5*(1), 119–137.

Bury, J. (2004). Livelihoods in transition: Transnational gold mining operations and local change in Cajamarca, Peru. *The Geographical Journal, 170*(1), 78–91. doi:10.1111/geoj.2004.170.issue-1

Bury, J. (2005). Mining mountains: Neoliberalism, land tenure, livelihoods, and the new Peruvian mining industry in Cajamarca. *Environment and Planning A, 37*, 221–239. doi:10.1068/a371

Buytaert, W., Célleri, R., De Bièvre, B., Cisneros, F., Wyseure, G., Deckers, J., & Hofstede, R. (2006). Human impact on the hydrology of the Andean páramos. *Earth-Science Reviews, 79*, 53–72. doi:10.1016/j.earscirev.2006.06.002

Buytaert, W., Zulkafli, Z., Grainger, S., Acosta, L., Alemie, T. C., Bastiaensen, J., ... Zhumanova, M. (2014). Citizen science in hydrology and water resources: Opportunities for knowledge generation, ecosystem service management, and sustainable development. *Frontiers in Earth Science, 2*, 26. doi:10.3389/feart.2014.00026

Cote, C. M., Moran, C. J., Cummings, J., & Ringwood, K. (2009). Developing a water accounting framework for the Australian minerals industry. *Mining Technology, 118*, 162–176. doi:10.1179/174328610X12682159814948

Falkenmark, M., Lindh, G., De Mare, L., & Widstrand, C. (1980). *Water and society: Conflicts in development. 2. Water conflicts and research priorities*. Oxford: Pergamon Press.

Fonseca, A., McAllister, M. L., & Fitzpatrick, P. (2013). Measuring what? A comparative anatomy of five mining sustainability frameworks. *Minerals Engineering, 46–47*, 180–186. doi:10.1016/j.mineng.2013.04.008

Garcia, O., & Gomez, C. A. (2006). *The economics of milk production in Cajamarca, Peru, with particular emphasis on small-scale producers* (PPLPI Working Paper No. 34), FAO A Living from Livestock - Pro Poor Livestock Policy Initiative. Cajamarca: FAO & IFCN.

Gelles, P., & Boelens, R. (2003). Water, community and identity: The politics of cultural and agricultural production in the Andes. In T. Salman & A. Zoomers (Eds.), *Imaging the Andes: Shifting margins of a marginal world* (pp. 123–144). Amsterdam: Aksant.

Instituto Nacional de Estadística e Informática - INEI. (2007). *Censos nacionales 2007: XI de Población y VI de Vivienda.* Lima: Instituto Nacional de Estadística e Informática.

Instituto Nacional de Estadística e Informática - INEI. (2012). *Uso consuntivo del agua superficial por vertiente (No. 2.22).* Lima: Instituto Nacional de Estadística e Informática.

Kemp, D., Bond, C. J., Franks, D. M., & Cote, C. (2010). Mining, water and human rights: Making the connection. *Journal of Cleaner Production, 18,* 1553–1562. doi:10.1016/j.jclepro.2010.06.008

Kuijk, F. (2015). *Water usage and efficiencies for irrigation in Northern Peru. A case study in Cajamarca, a region affected by mining industry* (Unpublished master's thesis). University of Leuven, Leuven, Belgium.

Li, F. (2009). Documenting accountability: Environmental impact assessment in a Peruvian mining project. *PoLAR: Political and Legal Anthropology Review, 32,* 218–236. doi:10.1111/plar.2009.32.issue-2

Mann, H. B. (1945). Nonparametric tests against trend. *Econometrica: Journal of the Econometric Society, 13,* 245–259. doi:10.2307/1907187

Moran, R. (2003). *Esquel, Argentina. Predictions and promises of a flawed environmental impact assessment.* Buenos Aires: Greenpeace Argentina Mineral Policy Center.

Mudd, G. M. (2007). Global trends in gold mining: Towards quantifying environmental and resource sustainability. *Resources Policy, 32*(1–2), 42–56. doi:10.1016/j.resourpol.2007.05.002

MYSRL. (2006). *Estudio de impacto ambiental suplementario Yanacocha Oeste.* Lima: Minera Yanacocha.

MYSRL. (2007). *Yanacocha: Balance social 2004 (No. COP 2939), Gerencia de Asuntos Externos y Comunicaciones.* Cajamarca: United Nations Global Compact.

MYSRL. (2008). *Reporte de sostenibilidad 2008.* Lima: Yanacocha.

MYSRL. (2009). *Reporte de sostenibilidad Yanacocha 2009.* Lima: Minera Yanacocha.

MYSRL. (2010a). *Reporte de sostenibilidad Yanacocha 2010.* Lima: Minera Yanacocha.

MYSRL. (2010b). *Estudio de impacto ambiental semi-detallado: Proyecto de exploración Maqui Maqui (Resumen Ejecutivo).* Cajamarca: Minera Yanacocha.

MYSRL. (2011a). *Gestion del agua en Yanacocha.* Lima: Minera Yanacocha.

MYSRL. (2011b). *Reporte de sostenibilidad Yanacocha 2011.* Lima: Minera Yanacocha.

MYSRL. (2012). *Reporte de sostenibilidad Yanacocha 2012.* Lima: Minera Yanacocha.

MYSRL. (2013). *Reporte de sostenibilidad Yanacocha 2013.* Lima: Minera Yanacocha.

MYSRL. (2014). *Reporte de sostenibilidad Yanacocha 2014.* Lima: Minera Yanacocha.

Newmont. (n.d.). Yanacocha-Peru. Retrieved from www.newmont.com/south-america

Observatory of Mining Conflicts in Peru. (n.d.). Retrieved from http://www.conflictosmineros.org.pe

Perreault, T. (2014). What kind of governance for what kind of equity? Towards a theorization of justice in water governance. *Water International, 39,* 233–245. doi:10.1080/02508060.2014.886843

Prado, E. (2012). Cajamarca soporta la escasez y el racionamiento de agua potable [News article]. Retrieved from http://larepublica.pe/13-01-2012/cajamarca-soporta-la-escasez-y-el-racionamiento-de-agua-potable

Rojas, P. (2010). *La gestión del agua en cuencas con minería. Limitaciones desde la sostenibilidad ambiental y la equidad social en la Sub Cuenca Porcón, Perú* (Unpublished master's thesis). Universidad Politécnica de Catalunya, Barcelona, Spain.

Scurrah, M. (2008). *Defendiendo derechos y promoviendo cambios: El estado, las empresas extractivas y las comunidades locales en el Perú.* Lima: Instituto de Estudios Peruanos.

SEDACAJ. (2012). *Plan Maestro Optimizado de la EPS SEDACAJ S.A. Periodo 2012-2017.* Cajamarca: Empresa Prestadora de Servicios de Saneamiento Cajamarca - SEDACAJ.

Servicio Nacional de Meteorología e Hidrología del Perú – SENAMHI. (2011). *Evaluación de las características hidrológicas en los Ríos Porcon y Grande de la Provincia de Cajamarca. Año Hidrológico 2010-2011*. Cajamarca: SENAMHI.

Sosa, M., & Zwarteveen, M. (2012). Exploring the politics of water grabbing: The case of large mining operations in the Peruvian Andes. *Water Alternatives, 5*, 360–375.

Sosa, M., & Zwarteveen, M. (2014). The institutional regulation of the sustainability of water resources within mining contexts: Accountability and plurality. *Current Opinion in Environmental Sustainability, 11*, 19–25. doi:10.1016/j.cosust.2014.09.013

Sperling, T., Freeze, R. A., Massmann, J., Smith, L., & James, B. (1992). Hydrogeological decision analysis: 3. Application to design of a groundwater control system at an open pit mine. *Groundwater, 30*, 376–389. doi:10.1111/gwat.1992.30.issue-3

Steel, G. (2013). Mining and tourism: Urban transformations in the intermediate cities of Cajamarca and Cusco, Peru. *Latin American Perspectives,40*, 237–249. doi:10.1177/0094582X12468866

Todd, D. K., & Mays, L. W. (2005). *Groundwater hydrology edition*. Hoboken, NJ: Wiley.

Vos, H. D., Boelens, R., & Bustamante, R. (2006). Formal law and local water control in the Andean region: A fiercely contested field. *International Journal of Water Resources Development, 22*(1), 37–48. doi:10.1080/07900620500405049

Yacoub, L. C. (2013). *Developing tools to evaluate the environmental status of Andean basins with mining activities* (Unpublished doctoral dissertation). Universitat Politècnica de Catalunya, Barcelona, Spain.

Yacoub, L. C., & Cortina, J. L. (2007). *Identificación y cuantificación de impactos medioambientales generados por MYSRL*. Barcelona, Spain: Universitat Politècnica de Catalunya.

Younger, P., & Wolkersdorfer, C. (2004). Mining impacts on the fresh water environment: Technical and managerial guidelines for catchment scale management. *Mine Water and the Environment, 23*, s2–s80. doi:10.1007/s10230-004-0028-0

Zwarteveen, M. Z., & Boelens, R. (2014). Defining, researching and struggling for water justice: Some conceptual building blocks for research and action. *Water International, 39*, 143–158. doi:10.1080/02508060.2014.891168

Disputes over land and water rights in gold mining: the case of Cerro de San Pedro, Mexico

Didi Stoltenborg and Rutgerd Boelens

ABSTRACT

This article analyzes different visions and positions in a conflict between the developer of an open-pit mine in Mexico and project opponents using the echelons of rights analysis framework, distinguishing four layers of dispute: contested resources; contents of rules and regulations; decision-making power; and discourses. Complexities in this study manifest how communities' land and water rights are circumvented by governmental bodies and ambivalent regulations favouring the large mining company. This process is importantly reinforced by international trade legislation. Multi-actor, multi-scale alliances may offer opportunities to foster environmental and social justice solutions.

Introduction

In 1996, Minera San Xavier (MSX), Mexican tributary of the Canadian mining company Newgold Inc., announced that it wanted to start a large open-pit gold and silver mine (Figure 1) in the municipality of Cerro de San Pedro, occupying 373 ha of community land. This was subject to great controversy as the scale and type of mining operation would put a heavy burden on the available land and water, not to mention adverse environmental effects. Resistance was fierce, and several opposition groups united themselves in the Frente Amplio Opositor (Broad Opposition Front, or BOF). Despite the opposition, MSX started operating in 2007. As of this writing, its presence is still being disputed.

Though mining is a highly profitable business for some actors, the downsides of mining activity are becoming more and more obvious. Environmental degradation, illegal land acquisition, water contamination, corruption, violence, resistance and conflict are often associated with mining development (Hogenboom, 2012; van der Sandt, 2009; Wilder & Romero-Lankao, 2006). *Campesino* (peasant) and indigenous communities are affected by mining activity in the area, and the livelihood strategies of mine-adjacent communities are often endangered through decreased access to and control over the land (Peace Brigades Internacionales, 2011; van der Sandt, 2009). Frequently, the economic 'benefits' promised by mining companies, for example in the form of

Figure 1. Before and after. On the left, the hill of Cerro de San Pedro before 2007, when operation of the mine started. On the right, the status of the landscape at the same time of year in 2013. The open-pit mine about 200 m from the centre of the village of Cerro de San Pedro has caused a large conflict that continues to date. (Left photo from BOF, 2013; right photo, Jesse Samaniego Leyva, 2013.)

temporary employment, do not outweigh the losses suffered (Perreault, 2014; Sosa & Zwarteveen, 2011; Yacoub, Duarte-Abadía, & Boelens, 2015). These negative effects often give rise to conflicts, and, unfortunately, conflict in a 'miningscape' is generally the rule rather than the exception. In 2013, the Observatorio de Conflictos Mineros en América Latina (OCMAL, 2014) registered 13 large-scale mining conflicts in Mexico, most of which involved foreign companies. One of these conflicts is taking place in Cerro de San Pedro.

This article elaborates how conflict arose over land and water rights between inhabitants of Cerro de San Pedro and MSX. The article examines how this 'natural resources conflict' is not just about rights to access resources, but also about underlying injustice in local, national and international rules and regulations, and about the question of the legitimacy and authority to shape these rules. It also shows how interconnected, powerful actor alliances, discourses, and knowledge claims profoundly influence the struggle over land and water in this municipality.

The article is based on literature and archival investigation throughout 2013 and field research in September–December 2013 in Cerro de San Pedro, with follow-up correspondence and conversations in 2014 and 2015. Semi-structured interviews were conducted with inhabitants of Cerro de San Pedro and surrounding villages, migrants who had left the zone, local municipality and government officials, mine representatives, mine-opposing groups, and journalists and scholars in San Luis Potosí. Next, a series of interviews were conducted (in three meetings each) with particular, representative individuals living in the mining area, whose life histories were compiled. Quotes and comments were taken from several interviews and conversations (all in Spanish and translated by the authors), which were conducted during both formal and informal encounters whilst performing fieldwork in Cerro de San Pedro (names of interviewees have been changed where needed to protect informants). The research for this article forms part of the activities of the international research and action alliance Justicia Hídrica/Water Justice.

The structure of the article is as follows. In the next section Mexico's neoliberal development, which paved the way for MSX to operate in Cerro de San Pedro, is elaborated. In the third section the background of the conflict over natural resources in

Cerro de San Pedro is discussed, after which the conceptual framework, echelons of rights analysis, used to analyze the conflict is explained. In the fifth section the results are presented. The concluding section reflects on how economic interests and associated discursive practices have had severe implications for the inhabitants of Cerro de San Pedro.

The background: Mexico, a protectionist state, takes a neoliberal path

To understand the conflict in Cerro de San Pedro, a brief look at the history of Mexico's laws concerning land, water and mining in the last century is essential. After the revolution of 1910, Mexico created a protectionist state in which land and water rights were a noncommodity. After years of unequal division of land and water under the *hacienda* system, the Mexican government expropriated the large landowners and reallocated the majority of the land and water to former day-labourers. These labourers formed farmer groups that collectively manage the resources to this day: the so-called *ejidos*, or social property sector. Under the *ejido* system, the majority of the allocated land is managed collectively whilst a small part can be cultivated for private purposes (Assies & Duhau, 2009). Under the law of *ejido* tenure, land was a non-negotiable resource. Article 74 of the Mexican Constitution stated that the ownership of common-use land is "imprescriptible, inalienable e inembargable [imprescriptible, inalienable and indefeasible]", i.e., land could not be transferred to third parties, land rights could not expire, and they could not be seized through an injunction (Herman, 2010). Water rights were linked to agricultural property rights under *ejidal* law, which means that they could not be sold, rented out, used on other lands, or used for purposes other than those stated in the grant (Assies, 2008).

However, after 1992, legislation on land and water rights changed. In the 1980s Mexico faced a severe economic crisis, and the World Bank, the International Monetary Fund and the Inter-American Development Bank demanded that Mexico adopt neoliberal policies if the country wanted to obtain international credit, similar to many other Latin American countries in the 1980s and 1990s (Achterhuis, Boelens, & Zwarteveen, 2010; Hogenboom, 1998, 2012; Wilder, 2010). The main focus of the restructuring of the economy was on opening the Mexican market to foreign investment. The social property sector and its regulatory framework of the time did not allow private ownership, as *ejidos* could not legally be privatized. This conflicted with the aim of increased foreign investment in Mexico, as land and water could not be converted into private and transferable commodities. Hence, according to neoliberal policy makers, if Mexico was to increase private (foreign) investment, legislation on land and water rights needed modification (Herman, 2010). Among others, the Agricultural Law, the Mining Law and the Foreign Investment Law were profoundly changed. To open up the mining sector to foreign mining companies, an amendment was made to Article 6 of the Mining Law that enables land to be alienated through "temporary occupancy". This provision enables mining activity to occupy land, and prioritizes mining above any other form of land use. The temporary occupancy permit is granted by Mexico's Ministry of Economics (Bricker, 2009; Herman, 2010). The 1992 market revisions paved the way for the North American Free Trade Agreement (NAFTA), which Mexico joined in 1994 (e.g. Hogenboom, 1998). Through NAFTA, foreign direct investment was stimulated greatly, and it was predicted that Mexico, as a developing country, would economically benefit most from these investments (Krueger, 1999; Ramirez, 2003). Meanwhile, for Canadian and US mining companies it became appealing to invest in Mexico due to the relatively low tax rates. It was shortly after the union with NAFTA that MSX announced its interest in exploiting the minerals in Cerro de San Pedro.

NAFTA has received criticism that environmental standards are easy to circumvent, due to the 'investor-state mechanism' that NAFTA encompasses. NAFTA aims to have investors of different countries treated equally and protected from expropriation by all levels of the (host) government. NAFTA's Chapter 11 gives an investor the right to challenge the government for noncompliance with the agreements made in NAFTA, in an international court, superseding national law. In the design phase, this mechanism was meant as a defensive measure to protect foreign companies against arbitrary and unreasonable government actions. However, it has had several deeply problematic side effects (Hogenboom, 1998, 2012; Solanes & Jouravlev, 2006, 2007). For example, first, it allows foreign companies to operate in the host country, but in case of a dispute they can go directly to the international arbitration process and entirely bypass domestic courts. Second, starting this process is relatively cheap and easy. This makes it an attractive option for foreign companies that wish to protect themselves against restrictions posed by new environmental laws or social security policies which could have a negative impact on their business (Mann & Von Moltke, 1999). The option of appealing to the international court under NAFTA is only available to companies operating under NAFTA, and not, for example, to communities or other non-business stakeholders who fear injustice, unequal competition, or socio-environmental costs (Herman, 2010; Nogales, 2002). On more than one occasion, multinational companies have used the possibility of suing governments for noncompliance with agreements made in NAFTA to prevail against environmental restrictions (Hogenboom, 1998, 2012; Solanes & Jouravlev, 2006, 2007). For example, near Cerro de San Pedro, in 2001, the American corporation Metalclad was awarded $16.5 million dollars by the NAFTA committee after the state of San Luis Potosí refused the installation of a hazardous waste transfer station. This shocked both national actors and international environmentalists (Kass & McCarrol, 2000).

The conflict in Cerro de San Pedro

Cerro de San Pedro has a long mining history. The gold and silver reserves in the area were already exploited by the indigenous inhabitants of Cerro de San Pedro, the Huachichiles, before the Spanish arrived. The Spanish conquistadores started exploiting the first mines in Cerro de San Pedro in the sixteenth century (Reygadas & Reyna Jiménez, 2008; Vargaz-Hernandez, 2006). Over time, several mining companies have come and gone in Cerro de San Pedro. Livelihoods were based on mining and agriculture, the latter practised for subsistence purposes under the *ejido* system.

The last large mining enterprise, before the current mining company, was active until 1948: the American Smelting and Refinery Company. After its shutdown, the majority of the miner families left the town to work in other mines in the north of Mexico; others went to San Luis Potosí in search of jobs. The remaining inhabitants developed new livelihood strategies, for instance based on tourism, making use of the local ecology and cultural heritage opportunities (interviews with local residents, October 2013; Reygadas & Reyna Jiménez, 2008; Vargaz-Hernandez, 2006). As soon as MSX announced that it wanted to exploit the minerals by means of an open-pit mine in 1996, opposition to the project started, involving inhabitants from Cerro de San Pedro and neighbouring villages, as well as relatives and ex-villagers now residing in cities like San Luis Potosí.

The mine covers 373 ha and consists of a large open pit, two waste dumps and a lixiviation area. In the lixiviation area, a water-cyanide solution is applied to the rock debris, dissolving the gold and silver particles. From the bottom of the heap the solution, now enriched with gold and silver particles, is drained. Eventually the water is evaporated, and what remains is a mixture of gold and silver known as doré. For this process, 16 tonnes of cyanide, dissolved in 32 million litres of water, is applied daily to the lixiviation area (Newgold Inc., 2009). MSX has water concessions for 1.3 MCM (million cubic metres) per year, but project opponents claim that the actual water extraction is much higher, and simple calculations show that the actual water needs of the mine are many times this figure (pers. comm., BOF member Eduardo da Silva, November 2014). For comparison, Interapas, which is responsible for the drinking water supply of the four municipalities close to the city of San Luis Potosí (more than a million people), has a total authorized annual extraction of 85 MCM of water (Peña & Herrera, 2008b). This use of cyanide, amongst other issues, has given rise to great opposition, as water is scarce in the area and cyanide is extremely toxic (Lutz, 2010). Years of litigation followed, during which a large number of court cases were filed, rejected, delayed or overruled by other courts. A large number of court cases were filed against MSX, questioning the legitimacy of the environmental/land use change permit, the mine's water use permit, and many other issues. Despite the mine's having lost a number of these court cases, MSX started its mining operation in 2007 (Herman, 2010; Peña & Herrera, 2008a). To date, MSX is still active in Cerro de San Pedro, but in 2015 the company announced it would gradually shut down the mine and accordingly start rehabilitation activities. These rehabilitation activities are described in MSX's project plan. They form part of a larger 'shutdown plan', in which MSX describes the process of closing the mine, and have been approved by the Secretaría de Medio Ambiente y Recursos Naturales (Secretariat of Environment and Natural Resources). Examples of planned rehabilitation activities are sterilization of the lixiviation area and reforestation of the mining pit. However, deep distrust of the mining company and the Mexican government, due to a lack of transparency and suspicion of corruption (as explained later), has caused the BOF to demand that the president of the municipality form a commission of independent experts to supervise the remediation work and its effectiveness (pers. comm., BOF, December 2015). Given the hugely problematic and controversial track record of the company, the government and other mining companies in the country in terms of (in)action regarding socio-environmental rehabilitation, this lack of trust is deep and generalized in the population.

Changing landscapes and waterscapes in Cerro de San Pedro

Unlike earlier decades' (tunnel-based) mining operations, the current open-pit mining practices have had a tremendous impact on the land and waterscape. The hill itself

(Cerro de San Pedro, the Hill of Saint Peter) – the place containing gold and silver particles – has been completely excavated (63 ha); beside it, two new hills of waste material have emerged (145 ha), and a newly constructed hill two kilometres to the south makes up the lixiviation area (120 ha). The new hills have altered the natural drainage pattern, blocking a dam and river in the village. Great amounts of dust cause severe pollution (Gordoa, 2011), and farmers in the area complain of crop failure caused by the pollution (interviews with local farmers, October 2013). The profound changes in the landscape caused by the mine were fiercely opposed. The litigation process in obtaining permits was seemingly never-ending, with courts referring to other courts and rejecting responsibility, creating a vicious cycle of court cases with no solid resolution (Herman, 2010; Peña & Herrera, 2008a). The differing opinions within the village drove a wedge between villagers, and a fully fledged conflict started in Cerro de San Pedro. Project opponents living in Cerro de San Pedro talk about cases of severe intimidation, aggression and violence against them, inflicted on them by both MSX employees and pro-MSX villagers. The economic interests in the realization of the mining project were enormous, and the national government put pressure on local authorities to issue the required permits. Oscár Loredo, the young mayor of Cerro de San Pedro, at first announced that he would not ratify municipal permits, but later changed his mind. He claimed that he was being put under great pressure by MSX, by the state, and even by the president (Vicente Fox), and that he could no longer stand the pressure. He claimed that he felt he had no choice and that his life was at risk. He added that his personal fears had made him change his mind about the matter. When challenged by one of the council members, the mayor responded: "Does my life not matter to you?" (Herman, 2010; Reygadas & Reyna Jiménez, 2008). Shortly afterwards, the municipal permits were ratified. A few years before, the former mayor, Baltazar Loredo (father of Oscár Loredo), was murdered after openly opposing the mining project (Vargaz-Hernandez, 2006).

Water availability in Cerro de San Pedro

Cerro de San Pedro and the city of San Luis Potosí are located in the hydrological watershed of the Valle de San Luis Potosí. This watershed stretches over approximately 1900 km^2 and supplies drinking water for about 90% of the San Luis Potosí population (more than a million people). The Valle de San Luis Potosí aquifer is being over-exploited: yearly, approximately 149.34 MCM is extracted from the deep aquifer, while only an estimated 78.1 MCM recharges it (Santacruz De Leon, 2008). As a way of mitigating aquifer over-exploitation, the Mexican government declared a *zona de veda* in the area.[1] *Vedas* are designed to prevent uncontrolled and unlimited water extraction from the deep aquifer, and thus aim to obtain a sustainable equilibrium, allowing human activities without degrading the environment (Conagua, 2012). Since 1961 the largest part of the Valle de San Luis Potosí aquifer has been subject to the *veda*, including the mine in Cerro de San Pedro (Conagua, 2004).

Another decree, issued 24 September 1993, designates 75% of the municipality of Cerro de San Pedro as a *zona de preservación de la vida silvestre* (zone for the preservation of wildlife). This decree was issued a few years before MSX announced its interest in exploiting the gold and silver reserves of the area (BOF, 2014). The State

Congress assigned Cerro de San Pedro and the surrounding area this protected status due to its ecological function and importance for watersheds. This was formalized in a state decree, which mandated that in 75% of the municipality of Cerro de San Pedro (1) no changes were to be made in the subsoil for a period of 20 years; (2) the area was not suited for industrial activities with high water consumption; and (3) it was acknowledged as having an important function for wildlife preservation (Gordoa, 2011; Vargaz-Hernandez, 2006).

Despite the *veda* and the watershed protection status, MSX has managed to bypass all regulations and court cases by obtaining a temporary occupancy permit and thus to start and continue mining. To illustrate the consequences, we present life histories from two affected families who took diverging paths. These stories show that some families outright opposed the transformation of their livelihoods, while others thought they could take advantage of the economic opportunities the mine would offer. After that, we present brief conceptual notes to examine the natural resource conflicts in Cerro de San Pedro.

Subsistence and opposition: the story of Doña Morena Sanchez Aguilar

"MSX arrived in 1996 and announced it wanted to start an open-pit mine. My husband and I were against the plans from the very beginning; we saw that they would destroy our surroundings, the land with which we are connected. My family descends from the indigenous Huachichiles: I belong here, this is my land. Moreover, we never saw the need to work for the mine. We saw another future for Cerro de San Pedro. Tourism was starting to pick up, and there were plans for the development of Cerro de San Pedro, such as the building of a hotel and restaurants. We didn't need the mine at all! Before MSX arrived, life was very different in Cerro de San Pedro. The town was united: we used to have dinner together, there were masses, sometimes we would dance on the square. This all changed when MSX came. They divided us. In the very beginning, almost everyone was against the plans of MSX. However, when MSX started to pay people for their 'vote', things changed. Our neighbours became our enemies! It got really violent, once they even tried to shoot my husband. The mine tried to silence the people who were against them. Houses were set on fire, our house was boarded up, all to intimidate us. Yet it was never proven that the mine was behind these things. These were really scary times, especially when our mayor was murdered. We stayed in the house and didn't meet up with our neighbours anymore. We were lonely. My husband, who was the fiercest in his opposition against the mine, passed away a year ago. Since then, the relationship with our neighbours has normalized a bit. We greet each other in the streets again. Yet we never interact with those who work for the mine. Sometimes MSX invites us for activities, but we never go. They might be starting to think we're now okay with the presence of the mine, but we are not. Moreover, we will never forget how our neighbours treated and threatened us a few years ago. This village is now a divided place. Life will never be the same here." (Interview with Doña Morena, November 2013)

Joining the opportunities offered by mining? The story of Don Vicente Estrada Diaz

"In 1996 Minera San Xavier came. They started going by our houses and told us that they were planning to start a mine here. They promised us that the mine would bring us a lot of benefits: job positions for everyone living in the village, a medical station, scholarships for our kids, and so on. I wanted a better future for my children; I wanted to give them the opportunity to study, which I never had. So to me this sounded like a great opportunity, and we agreed with the plans. My brothers and I sold about 19 hectares of our land, which we previously used to cultivate, to MSX. I am in favour of the mine out of necessity: I wanted job opportunities for me, my family and my village, and MSX was the only option we had. In other words, I am not in favour of the mine but in favour of a source of work. Of course I don't like the total change of our environment, or the contamination that mining activity brings about. Yet in MSX I saw the only way out of our poverty. Nowadays I doubt whether I should have been so positive about MSX in the beginning. Back then, I was one of the first to be in favour of the mine, and I convinced many of my neighbours. I really thought that MSX would give us a better life, yet did I know that MSX would not live up to all those promises they made? Yes, MSX improved our livelihoods, but not as much as we all expected and hoped. I feel I have let my people down: I was the one most positive about the mine, and look what we have now. We are all very worried about the contamination. The dust pollution and maybe the cyanide have severely affected our harvest. I still cultivate my fields, but the one located close to the lixiviation area hasn't given any yield at all. The plants are often covered in dust: how are they supposed to grow like that? It has rained quite a lot this year, and in other places my *milpa* grew pretty well. It must be the mine affecting my harvest, but what can I prove? I have no money or education to prove all this. We are very worried about the future. MSX is not going to operate here forever. What are we going to do when the mine is gone? The history is going to repeat itself. Everybody will leave our village, since there is no work. But this is where we were born, where we were raised and where we got married. How could we not love our land? Yet most of us stopped cultivating our land a long time ago. The best fields for cultivating were sold to MSX: now their offices are on top of it. Living off the land is a hard life and the young people are not attracted to this lifestyle anymore. On top of that, what can they do with a contaminated area? Only the old people, me, my wife and some other people, will stay here. But we have no option. When MSX goes, the life will go from our village as well." (Interviews with Don Vicente Estrada, October–November 2013)

Conceptual notes: examining the entwined layers of the natural resource management conflict

In mining areas such as Cerro de San Pedro, the use of natural resources such as land and water lies at the root of local inhabitants' livelihoods. Consequently, opposing interests and the struggle over access to and withdrawal of these natural resources has high potential for conflict. When new stakeholders, such as the MSX gold mine, enter the playing field and claim a substantial share of the resources, access rights are commonly reallocated through an interplay of (socio)legal, economic and political power. Redefinition of rights always causes some actors or use sectors to lose, such as less well-accommodated social groups, or the environment, while others reap the benefits and strengthen their political and economic positions – a basic tenet in political ecology (e.g. Forsyth, 2003; Neumann, 2005; Robbins, 2004).

In scrutinizing the conflicts that arise during the reallocation of natural resources, political ecology does not only attempt to focus on which population groups are most affected by these politics. It also aims to clarify the political forces that are at the roots of environmental distribution conflicts (e.g., Boelens, 2009, 2015; Brooks, Thompson, & El Fattal, 2007, 2013; Martínez-Alier, 2003; Neumann, 2005; Peet & Watts, 1996; Robbins, 2004; Turton, 2010; Wilder, 2010). Further, in order to grasp such political forces and their unequal outcomes in terms of resource distribution, there is also the need to focus on the ways in which environmental knowledge itself is produced, how 'knowers' are defined, by whom, and how they conceptualize 'environmental problems' and 'solutions'. All of these often implicitly generate uneven allocation of social costs and benefits. As Hajer (1993, p. 5) commented: "The new environmental conflict should not be conceptualized as a conflict over a predefined unequivocal problem with competing actors pro and con, but is to be seen as a complex and continuous struggle over the definition and the meaning of the environmental problem itself" (see also Forsyth, 2003). Different actors, with different socio-economic, cultural and political backgrounds, will commonly perceive and evaluate environmental transformations differently, and therefore use different frameworks to construct their "environmental imaginaries" (Peet & Watts, 1996, p. 37; see also Feindt & Oels, 2005).

As is common in such water extraction disputes, the conflict in Cerro de San Pedro exhibits many different levels and issues over which actors collide (Adler et al., 2007; Achterhuis et al., 2010; Brooks et al., 2007; Martínez-Alier, 2003; Turton, 2010; Wilder, 2010). To unravel the depths of the conflict, we use the echelons of rights analysis (ERA) framework (Boelens, 2009, 2015; Duarte-Abadía, Boelens, & Roa-Avendaño, 2015; Zwarteveen & Boelens, 2014). ERA can be applied to conceptually and empirically distinguish several mutually linked levels of abstraction within a natural resource management (e.g., water governance) conflict:

- The first echelon is about conflicts over access to and withdrawal of resources. In order to materialize these access and withdrawal rights, technological artefacts, infrastructure, labour and financial resources have to be in place. In this echelon the conflicts regarding access to and distribution of the resource(s) in question are examined.
- The second echelon refers to conflicts over the contents and meaning of the rules and regulations that are connected to resource distribution and management. Conflicts often occur over the contents of rules, norms and laws that determine the allocation and distribution of land, water and other territorial resources. Key elements of analysis in this field are the bundles of rights and obligations, roles and responsibilities of users, criteria for allocation based on the heterogeneous values and meanings given to the resource, and the diverse interpretations of fairness by different stakeholders.
- At the third echelon, conflicts over decision-making power are analyzed. Who is entitled to participate in questions about the division of land and water rights? Whose definitions, interests and priorities prevail? Who is able to exert formal or informal influence, and how?
- The fourth echelon is about the opposing discourses that are used by the different stakeholders to express the problems and solutions concerning land and water

rights. Different regimes of representation claim truth in different ways, and so legitimize their policies, plans, actions and the distribution of the resources. This last echelon seeks to coherently link all echelons together in one convincing framework (see e.g. Boelens, 2009, 2015; Zwarteveen & Boelens, 2014).

Throughout history and across continents and cultures, political and economic elites have often sought to justify their (often highly unequal) use of the environment by making use of a discourse that upholds this use as if it were for 'the greater good'. Opposing groups subsequently challenge these elite groups by forming their own counter-discourse. Hence, as the ERA framework illustrates, environmental conflicts do not concern only material practices but are simultaneously struggles over rules, over authority, and over meaning and ideological structures. In miningscapes, as in other arenas where actors fight over natural resources, rather than a search for absolute truths about environmental problems, we witness a battle about "the rules according to which the true and false are separated and specific effects of power are attached to the true", a struggle over "the status of truth and the economic and political role it plays" (Foucault, qtd. in Rabinow, 1991; see also Boelens, 2014; Boelens & Vos, 2012; Brooks et al., 2013; Forsyth, 2003; Turton & Funke, 2008; Wilder, 2010; Vos, Boelens, & Bustamante, 2006). As we see in this case about how the mining company MSX managed to get access to land and water rights in Cerro de San Pedro, discourses are not innocent tools: they often serve to justify particular policies and practices, and obliterate alternative modes of thinking and acting.

Unravelling the conflict in Cerro de San Pedro: the echelons of rights

Conflict over access to and withdrawal of land and water

In the analysis of the conflict in Cerro de San Pedro we see that on the surface the conflict revolves around access to land and water, the quality of these natural resources, and their use purposes and practices. Conflict over access to land is importantly expressed in the false lease contract that was presented by MSX, and the subsequent temporary occupancy of *ejido* land. Mexican law holds that the surface of the land belongs to the land right holders, in this case the *ejidatarios*, yet the subsoil remains the property of the government. This means that for MSX to obtain access to the land both a mining concession for the subsoil from the Mexican government and a rental agreement with the *ejidatarios* were required (Herman, 2010). Obtaining the mining concession from the government was not a problem. And since most rightful title-holders (*ejidatarios*) had left Cerro de San Pedro after 1948, MSX had the few remaining inhabitants (*avecindados*) sign a lease contract. However, these people did not hold the land title and thus could not legally rent the land to MSX. The false lease agreement was initially accepted, but eventually, in 2004, it was declared void. Meanwhile, between 1996 and 2004, despite the lack of a legal permit to access the land, MSX continued construction activities. Eventually, in 2005, a temporary occupancy permit was granted to MSX (Herman, 2010). The means by which the land was 'temporarily occupied' was also subject to controversy. In practice the occupancy means that local inhabitants are no longer able to use this land for agricultural purposes (for which the land was originally intended under the Agricultural Law), for tourism, or for artisanal mining.

Another part of the conflict is related to water use. MSX requires a large amount of water for operation of the mine, in an area in which water is already a scarce resource. The mining activity also has large impacts on the quality of the environment. Great controversy exists over the negative consequences for the quality of the land and water in the affected area, such as contamination of surface and groundwater; dust pollution; negative impacts on flora and fauna; contamination by heavy metals; and profound change of the surface of the landscape (Gordoa, 2011; Reyna Jiménez, 2009). The question is what quality of land and water will be left to local inhabitants after MSX closes the mine. Already evident is that, as explained above, the mining practices have transformed the territory into a huge open pit, altering its current and potential land uses, changing the village itself, drying out the groundwater wells, and even blocking the small river that once meandered through town but now has been usurped by the mine.

Disputing the content of the rules and regulations

Land rights. In Cerro de San Pedro, and Mexico in general, we see that the conflict is equally about the contents of the rules and regulations linked to mining and land and water use. At the very base of this conflict are two laws, the Agricultural Law and the Mining Law. Mexico's Mining Law considers mining to be of benefit to the entire society. Thus, any kind of exploration, exploitation and beneficiation of minerals should get preference over any other types of land use, including agriculture and housing (GAES Consultancy, 2007; Herman, 2010). However, this is not in accordance with Article 75 of Mexico's Agrarian Law, which states that "in cases where lands have been proven to be of use to the *ejido* population, the common land uses in which the *ejido* or *ejidatarios* participate may be prioritized" (Herman, 2010, p. 84). To ensure that mining activity can eventually take over all other forms of land use, Article 6 of the Mining Law enables land to be alienated through "temporary occupation" (Herman, 2010). Yet the Agricultural Law does not recognize this temporary-occupation instrument. Moreover, the Constitution considers land given to the *ejidos* "imprescribtible, inalienable e inembargable". Nevertheless, the denial of these fundamental rights is precisely what has taken place in Cerro de San Pedro. By denying *ejidatarios* the property of the subsoil as well as the surface, *ejidatarios* are legally being excluded from the game. The Agricultural Law recognizes them as the legal landowners, but the Mining Law considers mining to be in the public interest. So, the threat of having their land expropriated in the name of this public interest is ever-present for local villagers. If landowners do not agree to a lease contract they risk losing everything, with no compensation, through temporary occupancy. This puts them in an unequal negotiation position and forces them to accept unfair lease contracts (Clark, 2003; Ochoa, 2006). The Mining Law's temporary-occupancy provision de facto undermines the land titles of *ejidatarios*. Estrada and Hofbauer (2001) and the BOF state that local inhabitants are even further disadvantaged by the lack of legal follow-up: the Agrarian Attorney is obligated to supervise and assess the process of selling or renting *ejido* land to third parties, yet in practice this is often not done. In Cerro de San Pedro (and the majority of similar cases in which lease or sale contracts were produced between *ejidos* and a mining company), the *ejidatarios* were not informed about their rights and the possible risks of living close to mining activity.

In addition, as we elaborate below, San Pedro's customary rights and national agrarian laws supporting the position of Cerro de San Pedro's landowners are even further hollowed out by NAFTA rules, which stimulate and empower investors' rights and nullify the contents of local socio-legal arrangements protecting the environment and the community.

Water rights. Another subject of fierce conflict is the neoliberal policies that have converted water rights from a noncommodity into a tradable asset, which importantly favoured MSX's opportunities to operate in San Luis Potosí. These changes have allowed purchase and sale of presumably out-of-use water permits and the proliferation of well perforations in the *veda* zone (considered, under the new laws, 'relocation' of the old well), despite the clear objective of reducing over-exploitation of the aquifer. MSX obtained its water permits by making use of the new regulation and thus managed to buy 12 concessions totalling 1.3 MCM annually (Newgold Inc., 2009; Santacruz De Leon, 2008). Project opponents state that, keeping in mind the severe over-exploitation of the aquifer, tradable water rights put extra pressure on the aquifer in San Luis Potosí and endanger future water provision for San Luis Potosí inhabitants. Moreover, opponents claim that the granting of 1.3 MCM of a 'scarce resource' for mining purposes shows that the so-called scarcity is not an environmental condition, but rather the result of priorities that the government assigns to certain uses. They argue that the government decides that for some uses water is 'abundant' whereas for others it is 'scarce' (Peña & Herrera, 2008b). Scarcity in this sense is a social construction and political phenomenon rather than a natural state of the environment.

Besides the *veda*, the 1993 preservation of wildlife decree (preventing industrial water use) was also circumnavigated by MSX, triggering important conflict. In 2005 the *Sala Superior del Tribunal Federal de Justicia Fiscal y Administrativa* (Superior Chamber of the Federal Fiscal and Administrative Federal Tribunal) declared that the 1993 decree speaks of "industrial activity" (which has lower water rights prioritization), and since mining can be considered a "primary activity" (with higher priority), it is not subject to the decree (Herman, 2010). This decision provoked strong controversy, and project opponents objected in another court, starting a vicious cycle of court cases, seemingly without end. Project opponents claim that the location of the lixiviation area, in a zone designated for aquifer recharge according to this decree, besides being illegal, poses an extra threat of contamination of the aquifer (BOF, 2014).

Conflict over decision-making authority

As explained, MSX used a provision within the Mining Law that allows "temporary occupancy" of the land to acquire usufruct rights. This was granted by the Ministry of Economy in 2005, and thus the *ejidatarios'* rights to land and water were circumvented. Temporary occupancy has given birth to a much deeper discussion about the contents of the laws, how they interact and who has legal and/or legitimate power. In this case, the Mining Law was given preference over the Agricultural Law, but a large litigation process started, contesting the decision-making power of Mexico's courts: who is to decide whether the Mining Law supersedes Agricultural Law, or otherwise? This discussion is profoundly connected to power positions, discourse and knowledge, discussed in the fourth echelon.

Similar disputes relate to decision-making authority regarding the three decrees that have been issued in the past (*zona de monumentos, zona de veda* and *zona de preservación de vida silvestre*), which all have been overthrown in favour of mining activity. They were intended to protect the region socio-environmentally and culturally, but recent political power plays have altered and generated reinterpretation of the decrees with the aim of welcoming MSX to the area. However, the authority to over-throw these decrees (varying from the state governor to the national government) is fiercely disputed in the courts. BOF is actively fighting the decisions taken by autho-rities. BOF member Eduardo da Silva said, "Even when MSX leaves Cerro de San Pedro, our job is not done. There are so many other places in which the same thing is going on. We are not just questioning MSX, but equally the Mexican government: eventually, the government is the one who allows the law to be broken. Our goal is to change this governmental system, full of corruption, and to change the laws and the legal system that make it possible for companies such as MSX to operate in the illegal way they currently are" (pers. comm., October 2013).

The long legal battle and the different courts' rejecting responsibility and consequently referring to other courts have enabled MSX to continue operating while court cases remained pending. Several members of BOF mentioned that they feel that the Mexican government has deliberately adopted a 'from pillar to post' strategy of rejecting respon-sibility and referring to other courts, to postpone decision making and meanwhile give MSX the chance to operate (pers. comm., BOF member Eduardo da Silva, October 2013). Herman (2010, p. 85), quotes BOF's lawyer Esteban in her research on Cerro de San Pedro: "The legal processes are so poorly managed and the regulations are so vague that there are lots of ambiguities around the Agrarian Registry.... So the ejidatarios are not only against the mine, they're also litigating so that the courts recognize their rights."

International legislation has also put its mark on developments in Cerro de San Pedro, and brings up the question of which type of legislation (national or interna-tional) supersedes the other. Through NAFTA's Chapter 11, a foreign company may sue the host government, if it considers that the latter has not complied with agree-ments made under NAFTA, and thus put the company economically at a disadvantage. As UN's principal water lawyer, Miguel Solanes, writes about NAFTA and its threats to the public interest: "There is a tendency to replace the obligatory jurisdiction of the State with that of international arbitration tribunals.... Two types of economic players are thus created: those having all manner of guarantees, whatever the fluctuations in the economy, and those, usually ordinary citizens, who do not have any" (Solanes & Jouravlev, 2006, p. 63; see also Solanes & Jouravlev, 2007). In Mexican practice, on several occasions, local and national governments were sued by companies over the revocation or cancellation of environmental permits, after which the companies received large compensations for their economic loss from the host government (e.g. Kass & McCarrol, 2000). Solanes explains: "Only investors have legitimacy to request the intervention of investment arbitration courts, and to initiate suits and legal actions. They create the arbitration market, which depends on investors for its existence – the risk of capture and bias is strong. Since they are based on international agreements, investment courts trump national jurisdiction. In addition, other fora such as human rights courts lack the enforcement powers of the decisions of arbitration courts" (pers. comm., 20 December 2014).

MSX has threatened the Mexican government with NAFTA Chapter 11 to obtain the required permits at a time when the process seemed very difficult. Similar to the observation made by Warden and Jeremic (2007), just the threat of use of this provision has already caused a strong chilling effect in the case of Cerro de San Pedro. NAFTA provides an enormously powerful position for MSX vis-à-vis the national and local governmental authorities. Local communities are not allowed to object to resolutions taken within NAFTA, even though they often are the ones facing the greatest impact. Denying local inhabitants and communities the ability to file a complaint under NAFTA repudiates their legal status and stake in the conflict. As in Cerro de San Pedro, this creates enormous power differences between the local inhabitants versus the foreign company (Ochoa, 2006). As Miguel Solanes comments: "International investment agreements and their arbitration courts have made a travesty of local interests and power devolution. An arbitration court, at the international level, beyond local and national judges, ends up adjudicating conflicts between public local interests and global companies and investors. The international investment court does not only perform beyond local reach, but also outside the limits of public interest at the local level. Its mandate is to protect investors' interests, disregarding local problems" (pers. comm., 20 December 2014).

Conflicts among discourses

MSX's discourse is an essential tool in reaching its objectives, and helps understand how the issues explained in the previous echelons were tackled. The Cerro de San Pedro case witnesses how the mine's powerful discursive practices aim to morally, institutionally and politically legitimize their particular interests in using, managing and usurping the local natural resources, thereby arranging the human, the technological and the natural worlds in a 'convenient miningscape', as if these bonds were entirely natural.

Under its Corporate Social Responsibility programme, MSX claims that the company is deeply concerned with the environment, health, safety, and community development in both social and economic terms (Herman, 2010; Newgold Inc., 2012a). MSX states that it will provide jobs, education, healthcare and infrastructure to the local residents. On top of that, MSX claims to work with the newest techniques in order to minimize impact on the environment and reduce chances of contamination. Work and safety standards at work are said to be high; the wages that the mine offers would be high compared to Mexican standards (Newgold Inc., 2012b). The company's discourse explains that MSX is genuinely concerned with the livelihoods of its employees. By strongly advocating their commitment to security, health, environment and sustainability, MSX creates a discursive link between large-scale open-pit mining and positive development of the area. For example, MSX's annual Sustainability Report focuses largely on the job opportunities that MSX created for local inhabitants and on MSX's community development support, e.g. by means of the development of alternative income sources such as cactus nurseries, fish farms and a supply of microcredit for entrepreneurial initiatives to enable the people to sustain themselves after MSX abandons operation (pers. comm., MSX representative, November 2013; Newgold Inc., 2012b). By obtaining internationally recognized certificates confirming their 'sustainable operation strategy', such as the Conflict-Free Gold Certificate, MSX aims to comfort the public and government when it comes to health, society, environment and pollution (see also Vos & Boelens, 2014).

In many of its social activities – such as collective tree-planting days, the museum in Cerro de San Pedro in which the mining operation is explained and the 'benefit' for the local community is emphasized, and workshops on the production of silver jewellery – the company combines its strong power position with the creation of particular mine-convenient knowledge and facts, to make its mining truths become locally accepted truth. MSX has a very powerful position in this sense: it uses its economic position to influence public (e.g. mass-media) and governmental opinion, enhancing its social and political power position. Knowledge is actively created by MSX, as the company itself is in charge of monitoring the quality of water, air and soil. Thus, MSX establishes firm, triangular linkages between the three fundamental elements of Foucauldian discourse – power, knowledge and truth – mutually linked and shaping each other.

While some of the inhabitants have been convinced and have adopted the discourse of the mine's important economic, social and even environmental function for the region, others (for example those united in the BOF) have developed critiques and alternative or counter-discourses. BOF's mission is to stop MSX's activity in Cerro de San Pedro. To reach this, BOF actively spreads information in newspapers, social media and other outlets about the litigation process and the adverse environmental, cultural and economic effects caused by MSX, and organizes a yearly anti-mine music festival that takes place in Cerro de San Pedro.

Clearly, analyzing mining discourses and counter-discourses in Cerro de San Pedro gives insight into how the different groups perceive environmental problems and design solutions. Guthman (1997, p. 45) notes that the "production of environmental interventions is intimately connected to the production of environmental knowledge, both of which are intrinsically bound up with power relations. Therefore, the facts about environmental deterioration have become subordinate to the broader debates on the politics of resource use and sustainable development." Many villagers perceive that the process of knowledge production by the mining company, consultants, and state agencies reflects but also reinforces social and economic inequities in the area.

In everyday life, the discursive struggles and conflicts in the region – together with the diverging interests of villagers in relation to the mine's operations – have fuelled the divide-and-rule strategy that MSX has applied to the villagers since it arrived. When MSX came to Cerro de San Pedro, the village was unanimous in its objection to the mine. However, local inhabitants explain that the ambience in the village slowly changed and opinions on the mine started to diverge. For example, people say that certain families received money in return for their vote and others did not receive anything. The division between Cerro de San Pedro's villagers came to an all-time high when several pro-MSX villagers attacked some anti-MSX villagers, with the anti-mine villagers just able to run for their lives. Effectively objecting to the presence of MSX is more difficult for the anti-mine villagers if opinion is divided, an asset cleverly used by MSX.

Ways forward

Mining conflicts as in Cerro de San Pedro show a common feature, typical of most cases in Latin America as well as in many other regions: mining companies' power positions are reinforced by strong state backing and international investment agreement, producing a profoundly unequal negotiation position for the mining-affected populations. For the latter to obtain fairer and equal access to litigation possibilities, the mining company should be forced back to the negotiation table, and government institutions need to be made accountable for their key role as public service entities. To be successful, various cases provide evidence that this requires forging *multi-actor* alliances that work on *multi-scalar* levels, thus creating civil society networks that are internally complementary while connecting the local, national and global (see e.g. Bebbington, Humphreys-Bebbington, & Bury, 2010, and Boelens, 2008, for cases in Peru, Bolivia and Ecuador; Ochoa, 2006, for México; Urkidi, 2010, for Chile; and Hoogesteger, Boelens, & Baud, 2016, for Ecuador). By linking, for example, local village initiatives, women's groups, and journalists and newspapers with provincial indigenous and peasant federations, national ombudsman and civil rights offices, international research centres, and environmental and human rights NGOs, the negotiation forces (including access to research, information dissemination and possibilities for international arbitrage) can become more balanced and one-sided discourses can be challenged. Getches (2010) describes important opportunities for these multi-actor networks to use international norms and laws that can counterbalance powerful NAFTA-type agreements (see also Solanes & Jouravlev, 2007). Thus, besides more localized first-echelon resource struggles, in particular for second- and third-echelon strategies, the marginalized mining-affected communities may find important support through multi-actor and multi-scalar network action. These can seek to bend discriminatory rules or apply (inter)national protective regulations, and to balance the currently skewed decision-making powers. At the fourth echelon of ERA, such multi-actor network strategies will also strengthen the building of an alternative discursive framework, one that enables challenging the 'official' regimes of representation and can generate broader support for socially and environmentally friendly alternatives.

Currently the leading mining opposition group in Cerro de San Pedro, the BOF, is working on proposals for a new Mining Law, which builds on a more equitable and ecologically sound management of land and water resources. Getting the Mexican government to accept this will require great lobbying skills, a large network of influential partners and a well-balanced discourse. To this end, BOF is forging an alliance of local, national and international environmental organizations and universities, such as Pro San Luis Ecológico, Greenpeace México, and Amnesty International. These alliances provide BOF with access to new strategic-political opportunities, not only in Cerro de San Pedro but also in other mining arenas in Mexico. In the Cerro de San Pedro case, in which extraction activity is almost at its final stages, the main effort is now to try to reduce the damage done to the environment. Demanding ethically and ecologically responsible mining practices and waste cleanup, and also enabling alternative local livelihood opportunities, such as ecological and cultural tourism, might improve future job opportunities for the villagers whilst reducing the environmental impact.

Conclusions

Making use of the ERA framework shows that this mining conflict goes beyond the obvious struggle over accessing or defending land and water resources. In Cerro de San

Pedro, an exemplary struggle is being fought over land and water, yet with underlying struggles over the content of the rules and rights, and disputes regarding the decision-making authority to *make* those rules, which in the end seek to *distribute* the resources in particular ways. The discourses that are developed are not just weapons and counter-weapons in this struggle but also seek, in accordance with each party's interests and worldviews, to convincingly answer the questions and coherently link the issues raised in the first three echelons. In Cerro de San Pedro, they aim to depoliticize and naturalize MSX's miningscape, or, alternatively, show its profound contradictions as well as politically motivated 'mining truth', and arrange for 'alternative truths'.

MSX obtained access to the *ejido* land through the consent of Mexican government institutions which, despite the lack of required permits, allowed MSX to proceed with its operations on communal land. Moreover, the position of *ejidatarios* vis-à-vis the powerful mining company is further weakened by the systemic legal contradictions between the Agricultural Law and the Mining Law. The possibility of a temporary occupancy that can overrule the 'inalienable' land titles of *ejidos* inherently means that in Cerro de San Pedro, and throughout the country, *ejidos* can and will be pushed aside in favour of mining companies. Also, NAFTA has had great influence on the litigation process and on the bargaining position of local inhabitants. NAFTA's Chapter 11 gives foreign companies the opportunity to ignore national legislation concerning environmental and social rights and operate directly under the rules and regulations of NAFTA. NAFTA does not accept complaints from local inhabitants or communities, dismissing local communities' co-decision-making about their own futures.

The decision to overrule the existing decrees (the *veda*-regulations and the zone for the environmental preservation) in favour of MSX shows the Mexican government's eagerness for MSX to exploit the area. Although these decisions are contested by project opponents and remain to be decided in Mexico's courts, the circumvention of these decrees shows the extent to which an international, powerful actor like MSX can influence execution of national environmental legislation. Linked to these decrees is the granting of water concessions to MSX. Governmental allocation of 1.3 MCM annually to a mining industry contrasts with the total lack of water in a few neighbouring villages, with the endangered provision of water quantity and quality to a large city like San Luis Potosí, and with the official argument that in this valley water is generally an extremely scarce resource for which *veda* restrictions need to be obeyed. Water-scarcity declarations in the region clearly refer to political statements and priorities, which in the power context of San Luis Potosi easily bypass the natural state of this resource. The government declares a state of water scarcity when villages claim subsistence water use, yet it can simultaneously declare water abundance when a multinational mining company asks for large volumes of water to produce metals and a toxic environment. The economic interests of the few prevail over ensuring that the mine's neighbouring villages are provided with the most basic human right.

In the end, the changes in land and water rights in Cerro de San Pedro result from a complex interplay between different actors, where the court systems, officials and governments at diverse levels play a double and deeply troublesome role, and where multinational MSX has cleverly used the loopholes in the laws, plus its economic and discursive powers, to reach its objectives. In addition, international agreements as NAFTA have had a profound unethical impact on the litigation process, stimulating encroachment and sidelining social and environmental rights. The real victims of this interplay are the *ejidatarios* and inhabitants of Cerro de San Pedro, who have lost their alternative income-generating activities and access rights to land and water and who, once MSX abandons its operation, will be left without job opportunities, and with a polluted, entirely distorted environment.

Yet, the mine's deeply problematic impact may be reduced and halted in the near future. Through multi-actor networks that creatively engage in multi-scale action, mining-affected population groups, together with a variety of mutually complementary advocacy and policy actors, are striving to balance negotiation power and force the mining company to clean up mining residues and enable alternative local livelihood opportunities.

Note

1. *Zona de veda*: a specific area in which additional water use (on top of those concessions already granted) is not allowed; in short, new water concessions are no longer granted. This provision is designed to preserve the quantity and quality of the water, which can be either of superficial or subsurface nature (Conagua, 2012). However, the previously established amount of water for extraction may be so high as to prevent a healthy equilibrium from being reached. In the Valle de San Luis Potosí the *veda* refers to groundwater exploitation.

References

Achterhuis, H., Boelens, R., & Zwarteveen, M. (2010). Water property relations and modern policy regimes: Neoliberal utopia and the disempowerment of collective action. In R. Boelens, D. Getches, & A. Guevara-Gil (Eds.), *Out of the mainstream. water rights, politics and identity* (pp. 27–55). London: Earthscan.

Adler, R., Claassen, M., Godfrey, L., & Turton, A. R. (2007). Water, mining and waste: An historical and economic perspective on conflict management in South Africa. *The Economics of Peace and Security Journal, 2*(2), 32–41. doi:10.15355/epsj.2.2.33

Assies, W. (2008). Land tenure and tenure regimes in Mexico: An overview. *Journal of Agrarian Change, 8*(1), 33–63. doi:10.1111/j.1471-0366.2007.00162.x

Assies, W., & Duhau, E. (2009). Land tenure and tenure regimes in Mexico: An overview. In J. M. Ubink, A. J. Hoekema, & W. J. Assies (Eds.), *Legalising* land rights. *Local practices, state responses and tenure security in Africa, Asia and Latin America*. Leiden: Leiden University Press.

Bebbington, A., Humphreys-Bebbington, D., & Bury, J. (2010). Federating and defending: Water, territory and extraction in the Andes. In R. Boelens, D. Getches, & A. Guevara-Gil (Eds.), *Out of the mainstream. Water rights, politics and identity* (pp. 307–327). London: Earthscan.

Boelens, R. (2008). Water rights arenas in the Andes: Upscaling Networks to strengthen local water control. *Water Alternatives, 1*(1), 48-65.

Boelens, R. (2009). The politics of disciplining water rights. *Development and Change, 40*(2), 307–331. doi:10.1111/j.1467-7660.2009.01516.x

Boelens, R. (2014). Cultural politics and the hydrosocial cycle: Water, power and identity in the Andean highlands. *Geoforum, 57*, 234–247. doi:10.1016/j.geoforum.2013.02.008

Boelens, R. (2015). *Water, power and identity. The cultural politics of water in the andes*. London and Washington DC: Routledge/Earthscan.

Boelens, R., & Vos, J. (2012). The danger of naturalizing water policy concepts: Water productivity and efficiency discourses from field irrigation to virtual water trade. *Journal of Agricultural Water Management, 108*, 16–26. doi:10.1016/j.agwat.2011.06.013

BOF. (2013). *Website of BOF (Broad Opposition Front)*. Retrieved July 9th, 2013, from http:// faoantimsx.blogspot.mx/

BOF. (2014, April 5th). *Carta a Carmen Aristegui*. Retrieved October 23th, 2014, from http:// faoantimsx.blogspot.mx/2014/04/carta-carmen-aristegui.html

Bricker, K. (2009). Chiapas Anti-Mining Organizer Murdered. Retrieved April 8th, 2014, from http://narcosphere.narconews.com/notebook/kristin-bricker/2009/12/chiapas-anti-mining-organizer-murdered

Brooks, D. B., Thompson, L., & El Fattal, L. (2007). Water demand management in the Middle East and North Africa: Observations from the IDRC forums and lessons for the future. *Water International, 32*(2), 193–204. doi:10.1080/02508060708692200

Brooks, D. B., Trottier, J., & Doliner, L. (2013). Changing the nature of transboundary water agreements: The Israeli–Palestinian case. *Water International, 38*(6), 671–686. doi:10.1080/ 02508060.2013.810038

Clark, T. (2003). *Canadian mining companies in Latin America: Community rights and corporate responsibility*. Paper Conference CERLAC/York University and Mining Watch Canada, May 9 - 11, 2002, Toronto.

Conagua. (2004). *Registro publico de derechos de agua*. Retrieved March 25, 2014, from http:// www.conagua.gob.mx/Repda.aspx?n1=5&n2=37&n3=115

Conagua. (2012). *Vedas Superficiales*. Retrieved March 25, 2014, from http://www.conagua.gob. mx/ConsultaInformacion.aspx?n1=3&n2=63&n3=210&n0=1

Duarte-Abadía, B., Boelens, R., & Roa-Avendaño, T. (2015). Hydropower, encroachment and the repatterning of hydrosocial territory: The case of Hidrosogamoso in Colombia. *Human Organization, 74*(3), 243–254. doi:10.17730/0018-7259-74.3.243

Estrada, A. C., & Hofbauer, H. (2001). *Impactos de la inversión minera canadiense en México: Una primera aproximación*. México, DF: Fundar, Centro de Análisis e Investigación.

Feindt, P. H., & Oels, A. (2005). Does discourse matter? Discourse analysis in environmental policy making. *Journal of Environmental Policy & Planning, 7*(3), 161–173. doi:10.1080/ 15239080500339638

Forsyth, T. (2003). *Critical political ecology: The politics of environmental science*. London: Routledge.

GAES Consultancy. (2007). *Mexico - mexican market profile mining*. Ontario Ministry of Economic Development and Trade.

Getches, D. (2010). Using international law to assert indigenous water rights. In R. Boelens, D. Getches, & A. Guevara-Gil (Eds.), *Out of the mainstream. Water rights, politics and identity*. London: Earthscan.

Gordoa, S. E. M. (2011). *Conflictos socio-ambientales ocasionados por la minería de tajo a cielo abierto en Cerro de San Pedro, San Luis Potosí. (Licenciatura en Geografía)*. San Luis Potosí: Universidad Autónoma de San Luis Potosí.

Guthman, J. (1997). Representing crisis: The theory of himalayan environmental degradation and the project of development in Post-Rana Nepal. *Development and Change, 28*(1), 45–69. doi:10.1111/dech.1997.28.issue-1

Hajer, M. (1993). Discourse coalitions and the institutionalization of practice: The case of acid rain in Great Britain. In F. Fischer & J. Forester (Eds.), *The argumentative turn in policy analysis and planning* (pp. 43–76). Durham and London: Duke University Press.

Herman, T. (2010). *Extracting Consent or Engineering Support? An institutional ethnography of mining, "community support" and land acquisition in Cerro de San Pedro*. Mexico: Department of Studies in Policy and Practice, University of Victoria.

Hogenboom, B. (1998). *Mexico and the NAFTA environment debate. The transnational politics of economic integration*. Utrecht: International Books.

Hogenboom, B. (2012). Depoliticized and repoliticized minerals in Latin America. *Journal of Developing Societies, 28*(2), 133–158. doi:10.1177/0169796X12448755

Hoogesteger, J., Boelens, R., & Baud, M. (2016). Territorial pluralism: water users' multi-scalar struggles against state ordering in Ecuador's highlands. *Water International, 41*(1), 91–106. doi:10.1080/02508060.2016.1130910

Kass, S. L., & McCarrol, J. M. (2000). The Metalclad Decision Under NAFTA s Chapter 11. *New York Law Journal. Environmental Law*. Retrieved 27 April 2014, from http://www.clm.com/pubs/pub-990359_1.html

Krueger, A. O. (1999). *Trade creation and trade diversion under NAFTA*. National Bureau of Economic Research.

Lutz, D. (2010). Beware of the smell of bitter almonds. *Newsroom*. Retrieved July 23, 2014, from http://news.wustl.edu/news/Pages/20916.aspx

Mann, H., & Von Moltke, K. (1999). *NAFTA's Chapter 11 and the environment. Addressing the impacts of the investor-state process on the environment*. Winnipeg, Manitoba: International Institute for Sustainable Development.

Martínez-Alier, J. M. (2003). *The environmentalism of the poor: A study of ecological conflicts and valuation*. Cheltenham: Edward Elgar.

Neumann, R. P. (2005). *Making political ecology*. London: Hodder Arnold.

Newgold Inc. (2009). *Manifesto de impacto ambiental. Modalidad regional unidad minera Cerro de San Pedro - Operación y Desarrollo*. Cerro de San Pedro, San Luis Potosí: Minera San Xavier S.A.

Newgold Inc. (2012a). *Mining Project in Cerro de San Pedro*. Retrieved 9 July, 2013, from http://www.newgold.com/properties/operations/cerro-san-pedro/default.aspx

Newgold Inc. (2012b). *Reporte de Sustentabilidad 2012*. Cerro de San Pedro.

Nogales. (2002). The NAFTA environmental framework, Chapter 11 investment provisions, and the environment. *Annual Survey of International & Comparative Law*, 8(1), 6.

Ochoa, E. (2006). Canadian mining operations in Mexico. In L. North, T. D. Clark, & V. Patroni (Eds.), *Community rights and corporate responsibility: Canadian mining and oil companies in Latin America* (pp. 143–160). Toronto: Between the Lines.

OCMAL. (2014). *Conflictos Mineros en México*. Retrieved April 18, 2014, www.ocmal.com

Peace Brigades Internacionales. (2011). Undermining the Land. The defense of community rights and the environment in Mexico. *Mexico Project Newsletter*. London.

Peet, R., & Watts, M. (1996). *Liberation ecologies: Environment, development and social movements*. New York: Routledge Press.

Peña, F., & Herrera, E. (2008a). El litigio de Minera San Xavier: Una cronología. In M. C. Costero-Garbarino (Ed.), *Internacionalización económica, historia y conflicto ambiental en la minería. El caso de Minera San Xavier*. San Luis Potosí: COLSAN.

Peña, F., & Herrera, E. (2008b). Vocaciones y riesgos de un territorio en litigio. Actores, representaciones sociales y argumentos frente a la Minera San Xavier. In M. C. Costero-Garbarino (Ed.), *Internacionalización económica, historia y conflicto ambiental en la minería. El caso de Minera San Xavier*. San Luis Potosí: COLSAN.

Perreault, T. (ed.), (2014). *Minería, Agua y Justicia Social en los Andes. Experiencias Comparativas de Perú y Bolivia*. Justica Hídrica. Cusco: CBC.

Rabinow, P. (1991). *The foucault reader*. London: Penguin.

Ramirez, M. D. (2003). Mexico under NAFTA: A critical assessment. *The Quarterly Review of Economics and Finance*, 43(5), 863–892. doi:10.1016/S1062-9769(03)00052-8

Reygadas, P., & Reyna Jiménez, O. F. (2008). La batalla por San Luis: ¿El agua o el oro? La disputa argumentativa contra la Minera San Xavier. *Estudios Demograficos Y Urbanos*, 23(2 (68)), 299–331.

Reyna Jiménez, O. F. (2009). *Oro por cianuro: Arenas políticas y conflicto socioambiental en el caso Minera San Xavier en Cerro de San Pedro*. San Luis Potosí: COLSAN.

Robbins, P. (2004). *Political ecology: A critical introduction*. Chichester: Wiley.

Sandt, J. van der (2009). *Mining conflicts and indigenous Peoples in Guatemala*. The Hague: Cordaid.

Santacruz De Leon, G. (2008). La minería de oro como problema ambiental: El caso de Minera San Xavier. In M. C. Costero-Garbarino (Ed.), *Internacionalización económica, hierstoia y conficto ambiental en la minería. El caso de Minera San Xavier*. San Luis Potosí: COLSAN.

Solanes, M., & Jouravlev, A. (2006). *Water governance for development and sustainability*. Santiago: UN/ECLAC.

Solanes, M., & Jouravlev, A. (2007). *Revisiting privatization, foreign investment, international arbitration, and water*. Santiago: UN/ECLAC.

Sosa, M., & Zwarteveen, M. (2011). Acumulación a través del despojo: el caso de la gran minería en Cajamarca. In R. Boelens, L. Cremers, & M. Zwarteveen (Eds.), *Justicia Hídrica: Acumulación, Conflicto y Acción Social*. (pp. 381–392). Lima: IEP.

Turton, A. R. (2010). The politics of water and mining in South Africa. In K. Wegerich & J. Warner (Eds.), *The politics of water: A survey*. London: Routledge.

Turton, A. R., & Funke, N. (2008). Hydro-hegemony in the context of the orange river basin. *Water Policy, 10*(S2), 51–70. doi:10.2166/wp.2008.207

Urkidi, L. (2010). A glocal environmental movement against gold mining: Pascua-Lama in Chile. *Ecological Economics, 70*, 219–227. doi:10.1016/j.ecolecon.2010.05.004

Vargaz-Hernandez, J. G. (2006). *Cooperacion y conflicto entre empresas, comunidades, nuevos movimientos sociales y el papel del gobierno*. El caso de Cerro de San Pedro.

Vos, H. de, Boelens, R., & Bustamante, R. (2006). Formal law and local water control in the andean region: A fiercely contested field. *International Journal of Water Resources Development, 22*(1), 37–48. doi:10.1080/07900620500405049

Vos, J., & Boelens, R. (2014). Sustainability standards and the water question. *Development and Change, 45*(2), 205–230. doi:10.1111/dech.12083

Warden, R., & Jeremic, R. (2007). *The Cerro de San Pedro case. A clarion call for binding legislation of Canadian corporate activity abroad*. Toronto: KAIROS.

Wilder, M. (2010). Water governance in Mexico: Political and economic apertures and a shifting state-citizen relationship. *Ecology and Society, 15*(2), 22.

Wilder, M., & Romero-Lankao, P. (2006). Paradoxes of decentralization: Water reform and social implications in Mexico. *World Development, 34*(11), 1977–1995. doi:10.1016/j.worlddev.2005.11.026

Yacoub, C., Duarte-Abadía, B., & Boelens, R. (2015). *Agua y Ecología Política. El extractivismo en la agro-exportación, la minería y las hidroeléctricas en Latino América*. Justicia Hídrica. Quito: Abya-Yala.

Zwarteveen, M. Z., & Boelens, R. A. (2014). Defining, researching and struggling for water justice: Some conceptual building blocks for research and action. *Water International, 39*(2), 143–158. doi:10.1080/02508060.2014.891168

Mining and campesino engagement: an opportunity for integrated water resources management in Ancash, Peru

Robert Patrick and Lalita Bharadwaj

ABSTRACT

Mining has become Peru's largest source of revenue. There is evidence that many of the economic and social benefits of this burgeoning industry are not evenly shared across society. Uncertainty over water quality impacts from recent mining activity has been raised by indigenous *campesino* (peasant) communities in the Ancash Region highlands of central Peru. Adding to the growing conflict amongst competing water users is the current reduction of water availability caused by regional glacial recession. Based on interviews and focus groups this article explores opportunities for integrated water resources management to improve opportunities for campesino engagement in water resources decision making.

Introduction

Facilitated by international demand for precious metals and record-high global metal prices, mining now represents over 50% of Peru's national economy. Growing evidence suggests that many of the economic and social benefits of this recent and expanding industry are not evenly shared across Peruvian society. For example, it has been estimated that over half of Peru's indigenous population have been negatively affected in various ways by mining activities. Uncertainty over impacts on water quality from large-scale (formal) mining activity has raised human health concerns amongst *campesino* (peasant) communities in the Ancash Region of the Peruvian Andes (Figure 1). In addition to concerns over mining activity, other competing water uses include agriculture, hydropower and urban population growth. Adding further complexity is water storage losses from rapid glacial recession in the Andes as a result of climate change. Resource conflicts rarely exist in isolation (Mitchell, 2005), and this is certainly the case for the campesino communities of Ancash.

Powerless and with little voice, the indigenous people of Ancash symbolize those most at risk not only from potential mining impacts but from broader climatic and hydrologic change. In the face of such enormous uncertainty, this article explores how campesino communities may become more engaged in decision making and thus more able to meet their own needs. Using interviews and focus groups with campesino

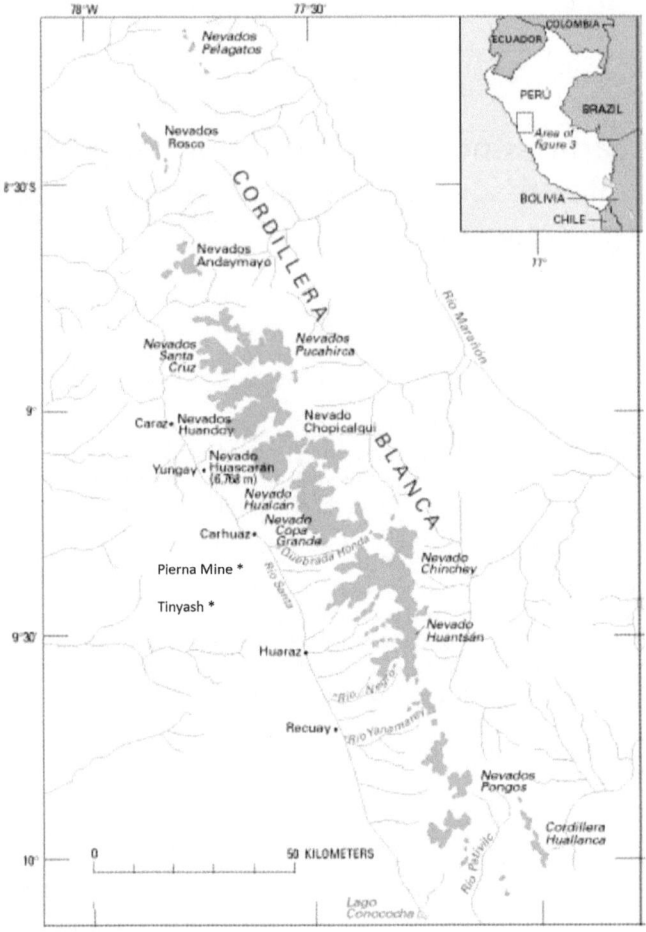

Figure 1. Ancash Region, Peru.

community members and other local actors, this article reports how the adoption of an integrated water resources management (IWRM) framework at the regional level has potential to improve the social, environmental and economic conditions of the indigenous population in the Ancash Region. While Peru is South America's most water-stressed country, it is also the country receiving the greatest investment in mine activity (Table 1). This physical and economic conundrum is infused with complexity and conflict when consideration is given to the social group most directly impacted by *in situ* mining: the indigenous peoples of the high Andes.

Arguably, the national government has prioritized the demands of external forces such as extractive industry, foreign enterprise and coastal population growth at the expense of highland, campesino communities (Galewski, 2010). In this article, this contested space is approached through exploratory research into how a water management regime, namely IWRM, may influence the outcome of industrial mining activities. Our hypothesis is that in the absence of a deliberate water management regime such as IWRM the outcome of any intensive land-based activity such as mining will contribute to uncertainty and conflict, particularly in the face of competing water users. Conflicts

Table 1. Ancash metal mining production, 2008.

Product	Peruvian producer (rank)	World producer (rank)	Production by region (metric tonnes)	Producing companies (Peru)
Copper	1	3	361,203	Antamina, Southern, Cerroverde
Lead	4	4	27,568	Volcan
Zinc	1	2	460,367	Antamina
Gold	5	6	12.57	Yanacocha, Barrick
Silver	2	1	531	Volcan, Ares, Buenaventura, Antamina

Source: after Galewski (2010).

have already erupted over water and mining in this region, and some of these conflicts have turned violent (Boelens, 2008). Given the national priority to promote mining expansion in the highland region of Ancash it is now prudent to explore whether IWRM may offer opportunities to both reduce conflict and allow greater inclusion of those people most directly affected by change and uncertainty regarding water resources.

This article reports opportunities for IWRM in a region of Peru experiencing rapid change in water availability and water quality and largely populated by people excluded from decision-making processes. More specifically, the objectives of this article are: (1) to assess the extent to which IWRM is currently practised at the local level; (2) to assess the potential benefits of adopting IWRM; and (3) to identify existing barriers to, and opportunities for, capacity building to support IWRM in campesino communities in the Ancash Region of Peru.

Integrated water resources management

The formal concept of IWRM can be traced to the Earth Summit held in Rio de Janeiro in 1992 (Biswas, 2004). Following this, the Global Water Partnership (GWP 2000) framed IWRM as "a process which promotes the coordinated development and management of water, land and related resources, in order to maximize the resultant economic and social welfare in an equitable manner without compromising the sustainability of vital ecosystems". Some have referred to the introduction of IWRM as a paradigm shift in the way water, and other resources, are to be managed (Saravanan, McDonald, & Mollinga, 2009). This paradigm shift was, in part, a response to what Bryant and Wilson (1998) describe as the failure of conventional environmental management as a "state-led technocratic problem-solving initiative devoid of understanding the political economic and cultural forces conditioning the process of environmental management". The early promise of IWRM was the focus towards the coordination, communication, governance and sustainability of water resources management and planning. This paradigm shift or 'integrated turn' in water resources management was not without its critics. The general thrust of the criticism was directed at how IWRM might be operationalized and at what scale (local, regional or national). For example, Mitchell (2005) questions how water and land-based systems are to be integrated for management purposes. Mitchell also pauses to consider what institutional arrangements would be required to break the current model of resource-based agencies working in relative isolation and with separate legislation and policy frameworks.

Biswas (2004) extends the critique, describing IWRM as a vague, indefinable, and un-implementable concept" and noting that the popularity of the concept greatly outweighs its "dismal" performance to "more efficiently manage macro- and meso-scale water policies, programs, and projects".

In contrast, others hold to the potential for IWRM to add value to water resources planning through increased collaboration and communication. Rather than replace past technological and positivist approaches to water management, IWRM introduces what Saravanan et al. (2009) describe as "communicative rationality as a place-based nexus for multiple actors to consensually and communicatively integrate decisions in a hydrological unit". Indeed, the United Nations has drawn attention to the benefits, indeed the necessity, of an integrated approach to water resources management to help overcome the "looming water crisis" in the less developed world (Falkenmark, Jägerskog, & Schneider, 2014). IWRM is built on the principles of multi-stakeholder participation in decision making, partnerships amongst key actors, holistic catchment-based governance, goal-oriented planning and coordination of land and water management initiatives (Mitchell, 2005). In the water resources literature IWRM is held out as a means towards local empowerment, shared decision making and appropriate technology to protect water quality and promote access to water supplies (Grigg, 2008).

While numerous operational definitions exist for IWRM, we refer to that provided by Grigg (2008, p. 5): "Integrated water resources management is a framework for planning, organizing and operating water systems to unify and balance the relevant views and goals of stakeholders." Implicit in this definition is recognition that IWRM is a guiding framework in which conflict and complexity amongst the competing interests of stakeholders may find balance.

Rather than a 'paradigm shift', the concept of IWRM may be viewed as offering a much-needed human and institutional dimension in water resources management and planning. The outcomes of IWRM will not always be easy to quantify. Yet, measured by improved relations across resource sectors, enhanced trust between disparate groups and better opportunities for public engagement, IWRM has much to offer. The IWRM framework was chosen for this study, in part, because Peru has adopted many of the policy prescriptions of IWRM, namely that water is an integral part of the ecosystem, a natural resource and a social and economic good. In Peru, water is the property of the federal state, and authority over water resources is situated in a single, national organization, the National Water Authority (Autoridad Nacional de Aguas, ANA). Throughout the country, ANA has identified 159 hydrologic water basins, each with a wide range of intra- and interbasin conflicts (Galewski, 2010).

The Ancash Region

The region of Ancash is centrally located in Peru, extending from the coastal area to the Andean highlands (Figure 1). The Santa River and its catchment is the major water-course flowing between two mountain ranges, the Cordillera Negra and Cordillera Blanca. The river originates in Lake Conococha (4000 metres above sea level) and flows 300 kilometres to the Pacific Ocean, draining an area of 12,000 km^2. The region's population was 1,063,459 in 2007, with 24% urban growth between 1993 and 2007.

Over this same period the rural population declined by over 6% (Galewski, 2010). Urban population growth in Ancash is centred on the inland capital of Huaraz as well as coastal cities and towns. These growing cities require a commensurate increase in water supply and service. Socio-economic stratification exists in Peru between coastal and highland areas. Highland communities are populated largely by indigenous people (campesinos), who are characterized by poverty, low literacy and education, disease, and a societal perception of ignorance. In contrast, the urbanized *mestizos* on the coast have better access to education, health services and wealth (Galewski, 2010). Highland communities face many infrastructure challenges, including electrical service, safe roads and water service. Compounding the challenges of urban water service and supply are commercial irrigation demands on the coast, and more recently, industrial mining activity in the highlands.

Superimposed on this socio-economic landscape are recent, and dramatic, changes in glacier mass balance in the Andes. The highland region's glacier melt water is an important input for crops during the dry season. Glacier recession is expected to significantly reduce water availability and threaten livelihoods in the highlands, including potable water supply but also campesino (subsistence) food security (Galewski, 2010).

Methodology

The methodology of the project embraced community-based participatory research in partnership with a local NGO, Asociación Urpichallay. Community meetings were arranged and included site visits to water infrastructure, campesino communities, and active mining areas. Asociación Urpichallay is a non-profit organization established in Ancash in 1991 to recover the value of Andean technologies and knowledge to improve the quality of life in rural areas. Asociación Urpichallay was invaluable to this research and facilitated access to the campesino communities. The national university in the Ancash capital, Huaraz, Universidad Nacional Santiago Antúnez de Mayolo (UNASAM), provided a vehicle and driver and also helped coordinate many of the meetings and field site visits in the region. Focus group discussions with local community members afforded opportunities to listen to individual experiences and for those experiences to be corroborated across the broader group. Interviews were conducted using a semi-structured questionnaire. Interviews and focus groups were conducted in the field in government, educational and organizational settings or in the campeseno communities. In all interviews and group meetings it was necessary to have a Spanish–English translator. In the campesino communities more than one translator was required, to translate from the local indigenous language, Quechua, to Spanish, and finally to English.

With support of our research partners, Asociación Urpichallay and UNASAM, community meetings were arranged with key informants. These included campesino communities, regional officials, academics, students and local water resource managers. A questionnaire was developed and provided to key informant representatives of these stakeholder groups as a means of gathering data to meet the research objectives.

With the assistance of representatives of our two partners we assembled a programme schedule for the duration of our visit in the Ancash Region. The activities in the programme included focus group discussions as well as site visits and one-on-one meetings with environmental committees in the Ancash communities of Marcara, Vicos, Tinyash and Shilla (Incapoo). These committees are composed of members of representative villages who work in partnership with Asociación Urpichallay to build community capacity in areas of environmental management and education and water management and monitoring.

Meetings were organized by UNASAM with non-profit organizations such as CARE International and Asociación Urpichallay. These non-profits work with campesino communities and their members to build capacity in the areas of environmental stewardship, water management and water security and education. Meetings were called with governmental and non-governmental organizations, including the Huascarán Working Group and Huascarán National Park. Huascarán National Park is in the Cordillera Blanca Range, in the Sierra Central of the Peruvian Andes. Representatives from Huascarán National Park are members of the Huascarán Working Group, a multi-stakeholder group including governmental agency representatives as well as NGO members whose mandate is to protect the interests and conservation of the national park and biosphere reserve.

A meeting was called to discuss impacts of climate change of glaciers impacting local campesino communities. This included a meeting with the chief engineer and coordinator of the national Department of Glaciology and Climate Change (Glaciologia-Climatologia-Gestion de Riesgos Procesos de Adaptacion al Cambio Climatico Obras Hidraulicas), Dr César A. Portocarrero Rodríguez. He and other members of the department conduct research on glacial retreat, management and monitoring and proactive engineering strategies and technologies to mitigate any impacts due to avalanches in relation to glacial lakes. A meeting was held with Chief Carlos Milla of the Federatión Agraria Departmental de Ancash, who represents the environmental, agricultural and cultural interests of 348 campesino communities of the region. There was a field visit with faculty from UNASAM to the headwaters of the Santa River and to Lake Conococha to observe lake level decline. Finally, a workshop was held with campesino environmental committee leaders as well as environmental promoters from the communities of Tumpa, Yungar, Shilla, Marcara, Vicos, Sanata and Catac.

In a three-week field research period in April 2011, 12 interviews and 4 focus group sessions were conducted with campesino community members and their leaders, academics, non-government actors and government officials. Local actors and agencies were relied on to identifiy the most relevant individuals and groups in the region for the interviews and focus group activity (snowball sampling). The local actors included local leaders, NGO representatives, farmers and local land users, as well as academics and a research scientist.

Overall, the above meetings and site visits provided opportunities to explore the extent to which IWRM is currently practised, and to determine capacity building necessary to adopt IWRM and any existing institutional barriers to, and opportunities for, capacity building to support IWRM in campesino communities. Information was also obtained on the ongoing and future water-related research projects campesino communities are conducting in partnership with Asociación Urpichallay and UNASAM. This was not undertaken as an exhaustive, high-level

Figure 2. Conceptual model of integrated water resources management.

policy analysis but rather as a more grounded assessment based on local practices, experiences and perceptions. As such, attention was focused on local actors and conditions through place-based inquiry, gathering qualitative data through key informant interviews and focus group activities.

To help achieve the second objective, to assess the potential benefits of adopting an IWRM approach, results from the initial assessment are compared and contrasted with a framework for IWRM derived from the literature (Figure 2). Specific interview questions looked for evidence of IWRM in the practices and experiences of local water managers and officials at the community level. In addition, questions explored capacity-building needs to support IWRM at the local level in the area of human, technical, institutional and financial capital.

For the purposes of this research, four defining features were drawn from the literature to characterize the IWRM approach: (1) catchment or river basin (rather than an administrative or political unit) as the management unit; (2) holistic water management framework, both institutionally and ecologically; (3) partnership and local cooperation; and (4) economic support and sustainability (Grigg, 2008; Mitchell, 2005). These four characteristics were used throughout this research to determine the presence or absence of IWRM in the Ancash Region.

Mining impacts

The principal impacts from mining in the highland regions are acid drainage, illegal waste dumping and mine tailings (Galewski, 2010). The highland region of Ancash is a major producer of copper, lead, zinc, gold and silver (Table 1). In 2012, mineral exports accounted for over 60% of Peru's export revenue (United States Geological Survey, 2014). Many of these mine operations date back many decades, pre-dating environmental standards as well as regulations set out under ANA. Documented studies show evidence of heavy metal drainage transported by rivers into the Pacific. This includes the Rio Santa, the largest drainage in Ancash. The largest gold and silver mine is owned by Barrick (Canada) and is just 10 km north of the region's main city, Huaraz. Pierna Mine opened in 1998 and has over 400 employees. Barrick built and controls a gated community in the city of Huaraz, complete with stores and a private company-controlled school along with housing for upper-level employ- ees. Antamina is a second mining company operating in Ancash and in other parts of Peru. The presence of these mining companies in Ancash is not without conflict. These mine operations are subject to strikes, protests and temporary closures. Conflicts and protest extend beyond the actual mine sites to include street protests in Huaraz and other urban and rural settings. Modern hydropower service is avail- able to the Barrick mine site, while the urban power supply in Huaraz and area is outdated and over capacity. A common complaint was the poor condition of urban infrastructure and lack of community contributions from companies like Barrick to improve local infrastructure such as roads. The larger discontent was over the negative impacts of mining on water and land, and thus human health. This was especially evident in the smaller highland regions, where campesino farmers and residents reported the degraded condition of surface water and soil. Stunted growth of vegetables and discoloured surface water were noted in the village of Tinyash. Local leaders from the village have been unable to talk with mine officials to express their concerns about the deteriorating conditions in their community, just down- slope of active mining operations.

Results

The results of the project are divided into four major themes based on the identified characteristics of IWRM noted above (Figure 1).

Catchment scale

Little evidence was found that water is managed on a catchment scale. Instead, water was reported to be managed through federal water law, with only limited regional government influence. The licensing and oversight of mining operations are similarly administered through federal law. Decision making around mine and water licensing, regulations and enforcement are removed from the Ancash Region and centralized in the national capital, Lima. As one participant noted, "There are two Perus: Lima and the rest of the country." Participants noted that there is no catchment-based authority to oversee water withdrawals, licensing or resource management. As a result, it was reported that no overall strategy is in place, no means of dispute resolution and no catchment-based agency overseeing the coordination of water use demands.

Holistic management framework

A holistic, or integrated, management framework is one that considers all factors affecting water supply and quality with a purpose of seeking solutions through good communication and coordination across sectors (Grigg, 2008; Mitchell, 2005). Such a model in Ancash would identify changes in highland water supply (climate change, deglaciation) and changes in water quality (mining, agriculture, urban wastewater). A more integrated management framework would help draw greater attention to the linkages between human health, food production and ecosystem services.

In contrast to this approach, single-sector interests such as international mining activities dominate local livelihoods and food security. Local fears were expressed over future water supplies as a result of climate change, combined with increasing competition for diminishing water supplies. Campesino communities are those most immediately affected by these changes and reported symptoms of the current fragmentation of water resources management in the region.

Partnerships and local participation

Campesino communities have no opportunity to engage in water management decisions. Federal water law fails to acknowledge highland campesino communities. As a result there is no local participation in water resource decisions. A centralized structure at the federal level maintains control over water licensing, with minimal devolution of authority to the regional government in Ancash. Campesino communities remain not only powerless under this regime but without opportunity to be heard.

Interviewees repeatedly stressed the lack of cooperation and communication with regional and national government officials. Partnerships involving campesino communities and mining interests or government are non-existent. Partnerships do exist with some NGOs, and there is hope for further partnership with UNASAM.

Economic sustainability

Appropriately priced water, following the IWRM approach, enables the collection of fees to support water infrastructure maintenance and upgrades (Mitchell, Priddle, Shrubsole, Veale, & Walters, 2014). Campesino communities are unable to collect fees under the existing water management structure and as such are under-resourced to maintain the water systems. In the absence of adequate and reliable financing, campesino communities continue to struggle to find disinfection products to maintain a safe water distribution system. As a result, highland drinking water systems remain vulnerable to contamination. Water quality testing was ad hoc across the visited communities. NGOs such as Asociación Urpichallay have provided assistance to some communities.

The long-term reliability of this assistance is unpredictable. In the communities visited, individuals volunteer to maintain the drinking water distribution system, accessing resources, such as sodium hypochlorite for disinfection, whenever possible. In other examples, the protection of potable water through riparian or reservoir fencing to restrict farm animal access is not feasible due to lack of material and financial resources.

In order for a water management issue to be listed in Table 2 the issue must have been raised during an interview or group discussion and corroborated at least once in another interview or group discussion.

Table 2. Water-related issues identified by participants in interviews and focus groups.

Most frequently mentioned
Reduced quality of water due to mining activities
Reduced health of local crops and human health from mine-contaminated water
Decreasing water supply from natural springs
Fragmented water management – national government makes decisions and agreements
Fear of inadequate water supply for their children and for future generations
Community-based water quality testing results not trusted by government
Campesino interests and concerns not taken seriously by government
No communication with campesino communities on land use or water management
Commonly mentioned
Increasing competition for existing water (e.g. mining, irrigation, coastal development)
Lack of funding to support local water treatment (e.g. chlorination)
Little opportunity for engagement in decisions across the three levels of government
Lack of coordinated, equitable development and management of water and land
No means of campesino participation in water management and planning
Lack of technical training and resources to undertake community-based water monitoring
Least commonly mentioned
Decreasing water supply due to deglaciation from climate warming
No opportunities for stakeholder forums to discuss water management
Lack of strategy to manage water on a local level
Lack of partnerships between water users (stakeholders) and campesinos

Source: the authors, 2016.

Discussion

In 2009 the national government approved a water law promoting integrated water management and established a regional catchment management council. Absent from this law is recognition of campesino populations in highland areas, much less any mechanism for campesino community participation in water management decisions. Water resources management in Ancash is fragmented and involves a tri-governmental system: the national government, the Ancash regional government and the municipal government. The Ancash regional government makes decisions involving land use, land use planning and water management, as well as urban and industrial development. The national government, more specifically ANA, is the primary authority responsible for water resources management. The national government has power to make and over-ride decisions of the regional and municipal governments that involve proposed industrial uses such as mining activity. For example, there are two levels of mining activity in the region: informal (small- and medium-operation coal mining) and formal (large-operation gold, silver and metal mining). The national government works with proposed large-scale formal mining activities in the area, independently of local govern-ments (regional and municipal). The Ancash regional government works with and handles decisions on proposed mining activities that are on a small-to-medium scale. The municipal government is not involved in decision making involving industrial water use and management but is involved in administration and management of water and wastewater infrastructure, health and education. To date, there remains no mechanism for campesino communities to engage in water resources management.

There is little in the way of coordinated, equitable development and management of water, land and related resources that would lead to a maximization of economic and social welfare in the region. IWRM approaches involve applying knowledge from various disciplines and insights from diverse stakeholders to devise and implement efficient, equitable and sustainable solutions to water and development problems. There is currently very little in the way of participatory planning for managing and developing water resources, or coordination or communication between the different water user groups and the decision makers of the various sectors, such as agriculture, mining and urban development.

Little evidence was seen of IWRM being practised at any scale of government, indicating disenfranchisement of campesino communities. It was not evident that venues for stakeholder forums to discuss water management had been available in the past or present, in either a formal or informal fashion. Lack of partnerships between water users (stakeholders) was evident in all field visits. Campesino communities manage water systems in isolation, with little if any financial or technical support from regional or national government. There was little evidence of communication across sectors, suggesting a severe limitation of opportunities for partnerships in managing water resources. A formal water management strategy, outlining goals and objectives for the management of highland water resources, was also absent.

Water security was the major issue identified in the highland areas of Ancash. Campesinos described decreasing water supply due to deglaciation from climate warming, reduced availability of ground-fed water and competing water usage (irrigation, mining, agriculture, forestry, increased human consumption and population growth). Mine contamination of surface water and related impacts to locally grown subsistence crops was noted. Lack of funding to support drinking water treatment (chlorination) was reported to be a major concern. Water treatment personnel act as volunteers in some campesino communities. In these situations, volunteers operate and maintain the water treatment and distribution system without pay and with minimal training. Members of the campesino communities regularly shared their fear that in the near future there will be inadequate water for their children and for future generations.

Capacity-building requirements for IWRM

Based on observations and discussions with key informant groups, four capacity-building requirements were identified as necessary for the adoption of a more integrated approach to water resources management in the area: financial, institutional, technical and human.

Financial capacity

There seems to be little financial support from regional, local or central governments to support inclusion of campesinos in water management. Financial support for water resources management primarily comes from non-profit organizations (e.g. Asociación Urpichallay). Normal support mechanisms from government seem to be absent and thus leave communities to collect user fees to support their local water systems. Local water fees are set at very low rates, considering the constituency, and in some instances go unpaid. For example, in the Shilla District the local water treatment operators did not have adequate supplies of chlorine to treat potable water for the community. It was unclear who was responsible for providing chlorine to the operators or if in fact the water treatment operators had to generate funds to purchase the chlorine. Financial capacity needs to be more strongly developed to support water management in the area. Consistent and sustainable funding is required to support effective and safe water management.

Long-term uncertainty of funding was evident in the current lack of adequate, sustainable and ongoing financial support from governmental institutions. It was observed that most support for actions to support IWRM is derived from NGOs (e.g. training, monitoring and management of potable water sources). In addition, it was observed that there are no mechanisms by which campesinos can access long- or short-term funding through the local, regional, municipal or federal governments to support IWRM. Furthermore, there are no respectful or beneficial relationships built with industry, and no industry motivation to fund or support IWRM; it appeared that the campesinos work in isolation (local institutional isolation) to support training and the monitoring and management of their water sources.

Institutional capacity

Institutional support for communities is provided primarily by non-profit organizations which provide soft funding. There is a lack of any defined senior government catchment authority to support local communities in initiatives of management, monitoring or treatment. There is also no regional government authority to support local initiatives; communities are left on their own, with little funding, training or education, to manage their water sources. Campesinos lack authority to regulate or enforce land-use regulations in highland areas they occupy. Adjacent land-use activities such as mining directly impact local water quality and quantity. Campesino community members frequently explained they have no voice and no power to influence change.

Senior governments hold authority for water management, and there is no institutional relationship between senior government and the campesino communities. This is a major institutional barrier to the adoption of an IWRM approach. Communities have no legal jurisdiction over land that involves their water source, and have no opportunities to comment on land-use development. Lack of legal mechanisms that provide voice to, or recognition of, communities was also observed as a primary barrier to IWRM adoption in the area.

Technical capacity

There is little evidence that campesino communities have access to training or advanced technology. For example, in one community (Tinyash) there were two hand-drawn community maps showing the main source of water, drainages and springs, location of houses, foliage, pastureland, and adjacent mining activity. Mine activity immediately upslope was reported to have altered surface water supply and quality. Impacts on water quality were visible in local subsistence food crops. Technical capacity building should be developed in the areas of surveying and mapmaking, water quality testing, soil testing and information sharing amongst stakeholders, including campesino communities.

The lack of access to educational opportunities to support technical training, as well as the absence of financial support to fund and provide opportunities for training at the local level, present barriers to IWRM. There was little evidence that employment opportunities or even paid positions for the technically advanced and trained community members exist or are available either inside or outside the local community. Community members also lack access to governmental or industrial trained technicians for consultative, advisory and training opportunities. Financial support from non-profit organizations such as Asociación Urpichallay is provided to advance technical training. These institutions provide a venue for workshops to support technical training (mainly for improved irrigation and water monitoring and treatment). They also support travel to workshops, arrange for speakers and provide meals during attendance. The financial sustainability of this, and other NGOs, is uncertain.

Human capacity

Some community members attend training sessions in water monitoring and management provided by Asociación Urpichallay. Community members who complete these training sessions become environmental leaders in their community and play key roles in water systems management. These people undertake maintenance, treatment and monitoring of the community water supply. Asociación Urpichallay also supports a two-year certificate programme in environmental promotion. Select community representatives who participated in the programme return to their communities as certified environmental promoters and support environmental stewardship activities. For example, trained environmental promoters may return to their communities and establish an environmental advisory group. This group would work with community members to identify and eventually develop community-based strategies to address an environmental issue in their community. Local community environmental promoters and established environmental committees meet regularly to share information and discuss issues of water resources and attend workshops supported by Asociación Urpichallay. Unfortunately, the federal government does not recognize local certifications, and water quality results from non-government approved laboratories go unrecognized.

Funding to support the work of NGOs such as Asociación Urpichallay is unpredictable. Sustaining and expanding these kinds of local programmes will require external funding support, combined with coordination with regional and federal government.

Opportunities for IWRM

There was evidence that partnerships are being built between UNASAM and the campesino communities. This early partnership was facilitated by Asociación Urpichallay and was mainly established to support technical capacity in the area of water monitoring and treatment. In addition, partnerships are being built between UNASAM and ANA. ANA is responsible for the design and implementation of sustainable water resources policies and irrigation nationally. The members of UNASAM and ANA could develop a framework to resolve the institutional, financial, technical and educational barriers to IWRM at the local level. Including the Ancash regional government in this partnership could lead to more financial support for local communities to build technical and human capacity for IWRM.

A catchment-based working group, encompassing all stakeholders (campesinos, land and water users, mining, agriculture, recreation, irrigation, health, education, social development and parks) should be formed. This would lead to better communication, trust and relationship building between all water user groups. The catchment working group would be responsible for developing an integrated water management strategy and work towards advancing education in schools on environmental stewardship, catchment education and the value of water from the perspectives of all user groups.

Conclusion

IWRM is broadly defined in the water resources literature as embracing a catchment-based unit of governance, multi-sectoral partnerships, multi-stakeholder collaboration and shared decision making, with both vertical and horizontal communication networks. The investigation indicates that IWRM is not practised in Ancash, to the immediate detriment of the natural environment and highland campesino communities. In addition, environmental impacts from international mining activity, as well as the cumulative impacts of other environmental factors, will only serve to compound the water management problems of the region. The present power differential between campesinos and the national government regarding water management may benefit international mining interests and the national government in the short term but will most certainly reduce water security in the medium and long term.

Mobilizing IWRM at the national level and scaling down to the regional and local levels will empower campesinos to join in water resources decision making. The immediate benefits of this arrangement would include local water quality monitoring, ability to properly fund water infrastructure, trust building between levels of government and improved campesino health. Longer-term benefits would include water and food security in Ancash, building adaptive capacity for climate and water availability change and shared decision making amongst all stakeholders. Key capacity-building requirements to support and maintain IWRM in Ancash have been identified. Building capacity at the local and regional levels, in concert with recent changes to federal water law, will put campesino communities in a better position to have greater voice in water resources management in Ancash.

Acknowledgements

The authors wish to thank the reviewers of an earlier draft of this article. Any errors or omissions are the sole responsibility of the authors. We are tremendously grateful to Asociación Urpichallay and UNASAM for their organizational support during the field component of this project. Finally, we acknowledge the generosity of the campesino communities, without whom this project would not have been possible.

References

Biswas, A. (2004). Integrated water resources management: A reassessment. *Water International*, *29*(2), 248–256. doi:10.1080/02508060408691775

Boelens, R. (2008). Water rights arenas in the Andes: Upscaling networks to strengthen local water control. *Water Alternatives*, *1*(1), 48–65.

Bryant Raymond, L., & Wilson, G. A. (1998). Rethinking environmental management. *Progress in Human Geography*, *22*(3), 321–343. doi:10.1191/030913298672031592

Falkenmark, M., Jägerskog, A., & Schneider, K. (2014). Overcoming the land–water Disconnect in water-scarce regions: Time for IWRM to go contemporary. *International Journal of Water Resources Development*, *30*(3), 391–408. doi:10.1080/07900627.2014.897157.

Galewski, N. 2010. *Campesino community participation in watershed management* (Unpublished Masters thesis). Georgia Institute of Technology, Atlanta, Georgia. Available from: https://smartech.gatech.edu/bitstream/handle/1853/34753/galewski_nancy_m_201008_master.pdf [Accessed 18 February 2016].

Global Water Partnership. (2000). Integrated water resources management. Technical background papers no. 4. Stockholm, Sweden: Global Water Partnership Technical Advisory Committee.

Grigg, N. (2008). Integrated water resources management: Balancing views and improving practice. *Water International*, *33*(3), 279–292. doi:10.1080/02508060802272820

Mitchell, B. (2005). Integrated water resource management, institutional arrangements, and land-use planning. *Environment and Planning A*, *37*(8), 1335–1352. doi:10.1068/a37224

Mitchell, B., Priddle, C., Shrubsole, D., Veale, B., & Walters, D. (2014). Integrated water resource management: Lessons from conservation authorities in Ontario, Canada. *International Journal of Water Resources Development*, *30*(3), 460–474. doi:10.1080/07900627.2013.876328

Saravanan, V. S., McDonald, G. T., & Mollinga, P. P. (2009). Critical review of integrated water resources management: Moving beyond polarised discourse. *Natural Resources Forum*, *33*, 76–-86. doi:10.1111/narf.2009.33.issue-1

United States Geological Survey. (2014). Glaciers of South America: Glaciers of Peru. Retrieved from http://pubs.usgs.gov/pp/p1386i/peru/occident.html

Questioning the effectiveness of planned conflict resolution strategies in water disputes between rural communities and mining companies in Peru

Milagros Sosa and Margreet Zwarteveen

ABSTRACT
Disputes between mining companies and surrounding communities over the access to, control of and distribution of water form an important part of the socio-environmental conflicts that large mining operations in Peru are producing. In order to mitigate environmental impacts, solve conflicts and deal with opposition to mining operations, governmental actors and mining companies make use of a combination of legal and technical strategies. This article questions the effectiveness of these strategies, focusing in particular on the longer-term sustainability of water resources, water-based ecosystems and livelihoods. Based on research carried out in the surroundings of the Yanacocha gold mine in Cajamarca, the article shows that although legal and technical conflict resolution strategies are effective in temporarily diffusing tensions, they do not address the underlying political causes of conflicts. Instead of these seemingly objective, neutral and quick solutions, the analysis suggests that solving environmental conflicts around large-scale mining operations requires explicitly admitting and dealing with the fact that these conflicts are always inherently political, situated, complex and power-laden.

Introduction

Since the 1990s mining activities have considerably intensified in the Andean regions of Peru, triggering a proliferation of socio-environmental conflicts (Bebbington, Humphreys Bebbington, & Bury, 2010; Bebbington, Hinojosa, Humphreys Bebbington, Burneo, & Warnaars, 2008). Between 2009 and 2012, more than 250 mining conflicts were reported. Many of these are about access to and control of land and water, and about the availability and sustainability of those resources for activities other than mining (Bebbington & Williams, 2008). In many cases, paradoxically, the communities that are articulating these concerns are often the same as those that expect and demand to benefit from the social and economic development opportunities that mining activities generate in rural areas (Bebbington, 2007). This paradox importantly co-shapes the nature and direction of conflict resolution scenarios: the fact that affected communities economically depend on mining companies limits their possibilities and willingness to express grievances and concerns. Often, conflicts happen within an institutional context that is not just characterized by large asymmetries in financial and political power, but that also separates the governance of mineral expansion from that of water resources and local development. Coupled with the political prioritization of mining development, this makes it difficult to effectively plan and regulate the interactions between mining operations and water.

Against this context, this article critically interrogates the effectiveness of planned attempts to solve or intervene in mining conflicts, particularly looking at whether and how these help ensure the sustainability of water resources, water-based ecosystems and livelihoods. We use selected examples of an iconic conflict, the one between the Yanacocha mining company and the rural communities of Combayo (Cajamarca, Peru), to show how conflict resolution typically happens by invoking the 'authority' of law and of science. The legitimacy of both importantly relies on their association with objectivity and neutrality. In other words, their strengths stems from the fact that they are designed to be non-political. We argue that the deeply political nature of mining conflicts makes it intrinsically difficult, or perhaps even impossible, to institutionally or scientifically safeguard objectivity and neutrality. This subsequently calls into question the effectiveness of the proposed technical and legal solution strategies. In particular, there is a danger that rather than correcting unsustainable or unjust behaviours, those in power – mainly the mining companies – use supposedly objective solutions as a legitimizing device to continue with business as usual.

We conclude that instead of trying to screen off socio-environmental mining conflicts from politics by resorting to science and law, it might be better to explicitly admit and subsequently deal with the fact that these conflicts are always inherently political and power-laden. That is to say, instead of basing solutions on forms of objective 'rightness', we suggest there is merit in acknowledging that they are always part of specific institutional contexts characterized by huge inequalities in voice and financial resources. This calls for much more emphasis on the process and power dimensions of environmental conflict resolution strategies.

This article is part of the first author's (M. Sosa) PhD dissertation. It is based on fieldwork done in Cajamarca, Peru, from 2008 to 2011. Like many farmers in the area of Combayo, during her fieldwork M. Sosa alternated living in the city of Cajamarca with stays in the rural town of Combayo. She also spent time in the communities of El Triunfo, Bellavista Alta, Bellavista Baja, Porvenir and Pabellón de Combayo. Information collection consisted of 40 semi-structured interviews in Cajamarca and Combayo with farmers, local and regional authorities, consultants, researchers, representatives of state agencies, the mining company and local NGOs. In addition, six in-depth interviews were held with presidents of irrigation canals and local authorities. M. Sosa also participated in community assemblies, communal work parties, herding activities, communities' parties, Sunday church services, cattle markets and judicial hearings. She used many of these occasions for observations, more or less formal interviews, oral stories, surveys, and focus group discussions, thus combining multiple research methods (Burawoy, 1998; Manson, 2002; Ranjit, 2011). Grey literature formed another important source of information.

After this introduction, the article presents the theoretical inspirations that informed the analysis. This is followed by background information on the case and a description of the conflict, and then an analysis of the legal and technical solutions that form important elements of strategies for dealing with opposition to mining operations. As a conclusion, the article discusses the effects of such strategies on rural communities, disputes over water resources influenced by mining operations, and the sustainability of water resources in general.

Water governance and socio-environmental conflicts

This analysis draws theoretical inspiration from an emerging literature about the contested nature of environmental problems on the one hand, and from scholarly attempts to come to grips with the many entanglements between science and society on the other. As for the first, scholarship that explicitly acknowledges the deeply contested character of environmental problems has a long history. Inspired by political ecology as a way to think about 'conflicts and struggles engendered by the forms of access to and control over resources and the inherent power relations in defining, controlling and managing nature' (Peluso & Watts, 2001, p. 25), this work assumes and posits linkages between the institutional regulation of property and the organization of society. In contrast to much conventional water-focused scholarship, it explicitly sets out to unravel how environmental planning and governance processes help reproduce social hierarchies and power relations. Hence, in political ecology terms, water governance is not just about water but also about the distribution of incomes, wealth and authority in society (Bridge & Perreault, 2009). This is one important reason why water is an intrinsically political and contested resource (Boelens, 2008; Mollinga, 2008; Panfichi & Coronel, 2010; Zwarteveen, Boelens, & Roth, 2005).

In line with this body of scholarly work, water governance can be defined as "the practices of coordination and decision making between different actors around contested water distributions" (Zwarteveen, 2015, p. 18). Such practices are thick with politics and culture, are linked to creative processes of imagining and producing collective environmental futures, and combine political problems of scale (spatial, ecological, administrative, temporal), with problems of coherence – the durable alignment of different people and different waters despite problems of incommensurability and political tensions (Bridge & Perreault, 2009). The implication of this perspective for water conflict resolution, the topic of this article, is that it can never be just a technocratic exercise, but should always engage with issues of (the organization of) power and politics.

Socio-environmental conflicts materialize when disagreements and contestations between different groups within society around natural resource (i.e. water) distributions, or the allocation of risks and hazards (Muradian, Martinez-Alier, & Correa, 2003), cannot be solved in a manner that is agreeable to all parties involved (Edmunds & Wollenberg, 2001). Such conflicts are symptoms of inadequate or ineffective political processes, as much as they signal problems of a more technical nature. This brings us to the second source of theoretical inspiration for this article, the literature on the entanglements between science and society. Because water is always contested, water (management and governance) questions cannot be resolved by just referring to objective, scientific information or analyses, but also involve matters of opinion and choice (Zwarteveen & Boelens, 2014) and have to do with interests and values (Muradian et al., 2003). Thus, to intervene in water conflict situations, scientific accounts of reality cannot be dealt with as an objective 'black box' separated from the context and from the political and social issues they are immersed in. On the contrary, pretensions of scientific objectivity or neutrality risk being purposively used to screen contentious questions off from explicit deliberation (Castro, 2007). As Li (2007) famously argued, questions that are rendered technical are simultaneously rendered non-political.

This article combines these insights about the intrinsically contested nature of water and the impossibility of separating politics from scientific or technical forms of knowledge as a framework to assess the effectiveness of conflict resolution strategies in mining areas in Peru. To summarize, we consider these conflicts more-than-technical in that they are indicative of wider power imbalances. These conflicts emerge when two or more actors or organizations compete for control of or access to water, and may evolve around issues of quantity, quality or opportunity (Pereyra Matsumoto, 2008). Urteaga (2011) thus aptly refers to these conflicts as expressions of political processes, with prevailing power relations co-shaping relations between the actors involved and their relations with water.

Background of the area and the conflict

The highlands of Combayo

Combayo, a rural town located in Cajamarca, has been the setting for a sequence of socio-environmental conflicts involving farmer communities and the Yanacocha gold-mining company as main actors. During the period of the *hacienda*, this rural town was known as the Hacienda Combayo. It formed one of the important estates in the northern region.[1] Today, about 13 of the 21 *caseríos* or communities of Combayo are located within the area under direct influence of Yanacocha mining operations (Yanacocha, 2007). Between 1992 and 1996, the company acquired about 4069 ha from 41 Combayo farming families, land that it needed to start operations in Cajamarca (Pascó-Font, Diez Hurtado, Damonte, Fort, & Salas, 2003). The combined effect of these sales and population growth is that at the time of this study, most land holdings in Combayo were small individual *minifundias*, ranging in size from 0.5 to 2 ha (INRENA, 2007). Here, like elsewhere in Cajamarca, the main economic activities are livestock and dairy production, together with some small-scale agriculture. For instance, surveys conducted in El Triunfo, one of the communities of Combayo, revealed that livelihood activities consisted mainly (76%) of small-scale farming and dairy production, with temporary employment at the mining site complementing families' incomes.[2] Farming and dairy production in Combayo rely on irrigation water that comes from streams or creeks fed by water from the Azufre River, a tributary of the Chonta River, which is part of a river basin of about 34,531 ha.[3] Assessments conducted during 2006 and 2007 by MINAG and sponsored by Yanacocha concluded that agricultural yields and dairy production in Combayo were lower than in other areas in Cajamarca Province. The study attributed this to low-quality seed, lack of proper soil fertilization, and inadequate farming practices. However, it also mentioned the lack of water for irrigation, or the lack of water security in the area, as a reason for low productivity (INRENA, 2007). A comparison of the water requirements of the existing farming systems in the watershed with water availability in the area reveals water shortages of about 38 MCM between May and October. Availability of water for irrigation is lowest and most critical in August (CEDEPAS, 2008), but water shortages can last for about eight months of the year.

Water is conveyed to the fields through a network of rudimentary canals. Some of these were constructed during the hacienda era, while others were dug more recently by farmers. Of the eight farmer canals in Combayo, about three were directly affected by the mining operations at the Carachugo site: Azufre Ahijadero (conveying about 100 L/s), Azufre Atunconga (150 L/s) and Azufre Ventanillas de Combayo (160 L/s), with a total of approximately 357 users and 885 ha of irrigated areas impacted (Mendoza Moreno, 2008). To manage the irrigation canals, farmers have organized themselves in water user associations, one *comité* for each canal. Some of them, like El Triunfo, hold official water rights given by the state. The associations are registered with the local water authority in Cajamarca[4] and led by a canal-president, who is responsible for distributing the available water supply, establishing the rotation schedule for the delivery turns, and organizing maintenance work. These canal-presidents also play a key role in mobilizing farmers in times of water disputes. They are the ones who speak for and represent the irrigators in the outside world. Their powers, means and resources to protect their canal's water rights are nevertheless limited (Sosa & Zwarteveen, 2012).

Before going into a more detailed description of the conflict, we briefly introduce the main features of the mining company. Yanacocha is a joint venture of the Newmont Mining company (USA), the Buenaventura mining company (Peru) and the International Finance Corporation, a member of the World Bank Group. It started operations in the region of Cajamarca in the 1990s. The mining concession of Yanacocha in Cajamarca consists of about 25,000 ha. The company operates a complex of open-pit mines, consisting among others of four leach pads and *in situ* processing facilities. Yanacocha's gold production for 2012 was 1.35 million ounces (Newmont 2012). Of all the mines operated by Newmont, Yanacocha is considered the most profitable (Bury, 2004). During the years of mining operations and because of its performance and expansion plans, Yanacocha has faced countless cases of socio-environmental conflicts (Arana, 2009; Bury, 2002; Deza, 2008; Guardia Nogales, 2011; Lingán, 2008; Sosa Landeo, 2012; Tanaka & Meléndez, 2009; Zavaleta, 2014) with several of the about 100 communities neighbouring its area of operations (Yanacocha, 2008).

Socio-environmental conflicts in Combayo

In 2005, farmers and authorities from Combayo started opposing the expansion of Yanacocha mining operations. They were particularly against the Carachugo II expansion project, an open pit, about 150 m in diameter and 180 m deep, in the high areas of Combayo. Yanacocha had obtained the authorization from the Regional Agricultural Authority (which also was responsible for water at that time) to use sources that were also used to supply water to Combayo. The mine had also obtained authorization to construct a dam on the Azufre River. *Comuneros* were concerned that the mine s uses and manipulations of water flows would negatively impact the quality and quantity of water available in the sources that fed this river, which in turn would have implications for their irrigation water.[5] In particular, three lakes were going to be compromised: Corazón, Patos and Estación 1.

The farmers and communal authorities mobilized to launch a collective complaint to the Agriculture Authority in Cajamarca against the authorization given to Yanacocha. In support of the claims of the population, the Agriculture Authority agreed to revisit its authorization. Yanacocha, fearing obstruction of its plans, reacted by engaging in conversations and negotiations with some Combayo representatives. The mining company succeeded in reaching an agreement with these representatives, in which it promised to protect the water sources of Combayo.

In spite of this agreement, however, communities of Combayo in the Azufre Watershed, and particularly in the area of direct influence of Yanacocha, began noticing changes in the water flows in their canals after Yanacocha had started its operations: "the water was different".[6] After they irrigated their grass it changed colour – "it got yellowish" – and animals that drank this water got sick. They also noted a reduction of water flows. As Yanacocha's operations happen in the upstream areas of the Azufre River, near the three lakes, *comuneros* did not hesitate to attribute the reported changes to the mining operations. Farmers' discontent and anger with the company were also fed by the collapse of the dam that Yanacocha was building on the Azufre River; this damaged farmers' plots and crops.[7]

Led by the presidents of the water user associations, irrigators decided to stage an organized campaign to demand Yanacocha's compliance with the promises it had made in 2005. In the first days of August 2006, around 600 people from the affected areas (Bellavista Baja, Bellavista Alta, El Triunfo, Porvenir and Pabellón de Combayo) headed for two of the Yanacocha mining sites – the Chaquicocha open pit and the Carachugo mountain – to protest (Figure 1). They were repulsed by the security company of Yanacocha and the police officers the company had hired to protect the mining site, who used tear gas and guns to stop the protesters.[8] In the confrontation, Isidro Llanos, a farmer from the community of El Triunfo, was hit by one of the bullets and died. During the ensuing turmoil, two workers of Yanacocha were taken by the farmers. Yanacocha interpreted this as a kidnapping, and held two farmer leaders responsible.[9]

Right after the protest, the mayor of Combayo joined with some other authorities in attempts to arrange meetings between the farmers and Yanacocha to discuss the impacts of its operations on water. According to them, however, all their initiatives were unsuccessful. This is why they decided to revert to less peaceful means. For about 20 days, they blocked the main access road to the mining site, preventing Yanacocha from operating as usual. In the media, representatives of the company stated, "Because of this conflict, Yanacocha has decided to stop operations at the expansion project."[10] The vice president of Newmont in Latin America, Carlos Santa Cruz, announced that this would represent a loss of about USD 700,000 for the Peruvian state and about USD 2 million for the company.

The continuous tensions in Combayo, the blocking of the road, and the company's announcement that it would stop its mining operations in the region aroused the attention of the central government. Even the prime minister, Jorge Del Castillo,

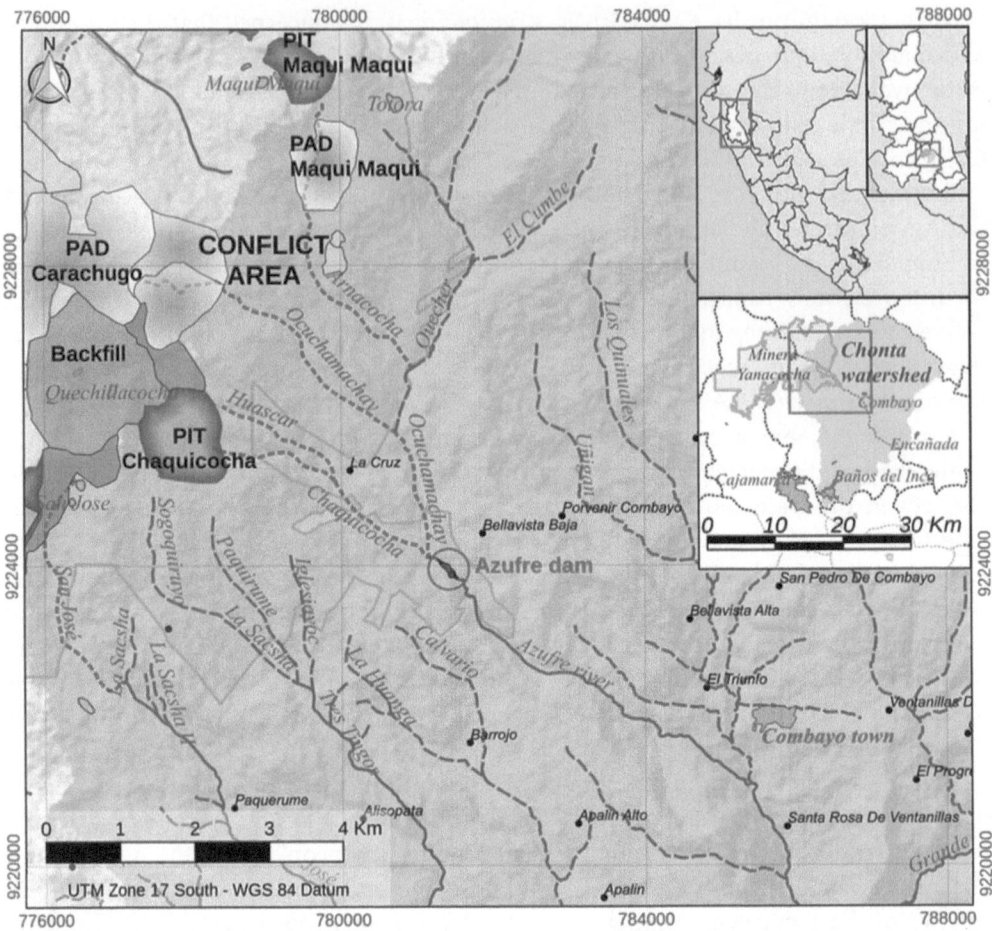

Figure 1. The conflict area in Combayo, Cajamarca, Peru. The figure shows the conflict area in the vicinity of the Yanacocha mine, the Chaquicocha and Ocucho Machay creeks, the Azufre river and the dam. The box in the upper right corner zooms in on the conflict area, the middle box shows the size of the Chonta watershed and the Yanacocha mine site in relation to the town centre of Combayo and the city of Cajamarca. Source: Designed by C. Cerdán based on ZEE-OT Cajamarca: Gobierno Regional de Cajamarca (2011), Zonificación Ecológica y Económica base para el Ordenamiento Territorial del departamento de Cajamarca, retrieved from http://zeeot.regioncajamarca.gob.pe/ sites/default/files/DocumentoZEEfinal.pdf; M. Salazar, El camino del Azufre, La República, 3 September 2006, retrieved from http://larepublica.pe/03-09-2006/el-camino-del-rio-azufre.

intervened. He decided to personally help solve the conflicts by mediating between the farmers and the company (Villar, Gonzales, & Roncal, 2006). His intervention resulted in a public meeting in September 2006, one month after the conflict, in which the Peruvian government, the company and the authorities of Combayo signed an agreement, called the Acta de Combayo.[11] This agreement consisted of a whole menu of solutions to the water problems in the area, ranging from assessments and management plans to promises of work and investment projects. In addition to those agreements, the meeting was also used to explicitly and formally obtain Combayo's promise that it would stop opposing the development of mining operations in the area.

Satisfied by this outcome, Yanacocha stated that dialogue is the only way to understanding and development. "What started as a conflict ended with an agreement for mutual support, inclusive dialogue and long-term development plans" (http://www. yanacocha.com/informes-especiales/, accessed 2009). The farmers, however, were less satisfied. Especially those from the affected communities were suspicious about the outcomes of the negotiations.

The following sections present a more detailed analysis of these negotiations. The analysis shows that the conflict resolution strategies favoured by the mining company and state officers importantly rely on legal and technical (or scientific) forms of authority. As noted, the conflicts took place in a context of large financial and political power asymmetries. Indicative of this is that Yanacocha had agreed to make 'social' investments of about USD 1 million in 2007 and 2008 in Combayo. The communities that (hoped to) benefit from this support (with sprinkler irrigation systems) were also the ones that complained that their water sources were deteriorating because of the mining operations.[12] The economic dependence of the communities on the company obviously weakened their bargaining power, and negatively affected their ability to hold the company accountable for its water actions.

Invoking legality: operating according to law

For the mining company (often in combination with some government actors), a first important and powerful strategy to deal with communities' complaints and reduce tensions is to convince all involved that everything they do is within the law. The reasoning is that if something is legal, it must be right, even if it is clear that this legal rightness says little to nothing about its social or environmental integrity.

Mineral expansion projects like Yanacocha's Carachugo project have to comply with more than a dozen regulations and norms to get a license to operate from the state (Torres, 2007). A large number of government authorities, such as the ministries of energy and mining (MEM), agriculture (MINAG), and environment, as well as the local and the national water authorities, have thus been involved in granting Yanacocha permission to operate. Permissions were given, among others, for accessing and collecting water, as well as for its extraction, management, treatment and disposal (MWH, 2012). Also, as part of the requirements established by the Peruvian environmental legislation, environmental impact assessments (EIAs) have to be publicly presented and discussed.[13] For the Carachugo expansion project, this took place in meetings organized by the MEM in Cajamarca in 2003 and 2004. These meetings happened with the help of private consulting companies, who informed the general audience of the actions that Yanacocha would be developing in the area. After these public hearings in Combayo (December 2003) and in Cajamarca (January 2004), the EIA was approved. The MEM considers these public hearings to be processes of consultation and public participation. The mere fact of their being held is enough to comply with the law. However, it is debatable whether these meetings are effective in terms of communication.[14] As is also shown below, many people from Combayo – authorities as well as farmers – were not properly informed regarding how mining operations would affect their water resources.

In 2008, Yanacocha wanted to develop mining operations at the Carachugo Mountain, compromising the areas of the Ocucha Machay and Chaquicocha Creeks, tributaries of the Azufre River. These operations, as well as previous hydrological and hydrogeological studies done by Yanacocha, were authorized by the water authority and the MINAG through administrative resolutions 051-2008-INRENA-IRH and 367-2008-INRENA-IRH. These stated that Yanacocha was authorized to execute "surface draining works at the influenced area of the Los Patos, Corazón and Estación 1 Lakes at the Ocucha Machay Creek to facilitate the expansion of the ... leaching pad at the Carachugo Project".[15] Having all the authorizations made it seem as if everything was agreed for the company to proceed. Yet, not all were in favour of the mining company's plans. The deputy governor of Combayo, for one, complained against Yanacocha, arguing that the people of Combayo had not been informed about these works. They were particularly upset about the fact that they had not given any authorization to the company to proceed with drying out Combayo's water sources.

According to the deputy governor, the EIA did not contain any clear reference to or information about the removal of lakes. The only answer from the representatives of Yanacocha to the complaints was that they had duly complied with all the legal requirements, met all the regulations and had obtained all the permissions needed to proceed since the approval of the EIA in 2004, a process that was validated by the local authorities of that time. They also referred to the authorization given by the National Water Authority resolution (367-2008-INRENA-IRH) approving hydrological and hydrogeological studies and drainage plans in the area of the lakes.

Besides letters to the local representatives of Yanacocha, the deputy governor issued a letter to the highest representative of Newmont in Latin America.[16] In the letters, he expressed his discomfort regarding the lack of communication from the company about the drying out of the lakes. He mentioned that those actions were not communicated to the population and that when he had asked for explanations, the response was merely that they "are not doing any work that does not have all the permissions of pertinent authorities and with the full acknowledgment of the population". Arrogantly, the company added that they could do as they pleased within their concession. According to the deputy governor, if there was an authorization given by the population, this must have been given in a dubious way, without those granting the authorization realizing what they were doing.[17]

In addition to permissions or authorizations given by the state, the law also requires that companies get permission to operate from the communities and their authorities. How this community approval should be obtained, however, remains vague and questionable (Li, 2009). Commenting on communities approving documents and actions, a Cajamarca regional officer told the story of the leaders of Combayo signing a document thinking that it was a simple request for a regular inspection of the headwater areas to be done by the Cajamarca Water Authority (ATDRC). They did not realize that the document would be used as an authorization to dry out the lakes: "The signatures [of the authorities] were for the inspection of the lake area, not for making them disappear!"[18]

As part of the agreement signed by the state, Yanacocha and Combayo, the company promised not to make any more legal accusations against the farmers. The farmers and communities in turn had to state that they would not oppose mining operations in the area. However, the farmers from the affected communities that organized the protest commented that little was achieved for their areas in terms of environmental conservation, water protection and socio-economic improvements. According to them, they came out of the conflict worse off than they had entered it: with the death of Isidro Llanos, and the anxiety provoked by the judicial case that followed the conflict. "The people fear that they will be judicially denounced by the [company]; the judicial processes have restricted people."[19]

Invoking science: technical knowledge supporting operations

A second important strategy of Yanacocha in alliance with the Peruvian state to solve conflicts is to rely on scientific expertise to produce supposedly objective assessments of how mining will affect the quantity and quality of water flows.

As mentioned earlier in the article, small-scale agriculture, livestock and dairy production constitute the permanent livelihood activities of rural households in Combayo. Those activities crucially depend on the availability of water in the canal networks that are fed by water from creeks and the Azufre River. That MINAG and Yanacocha are well aware of the criticality of local water resources for sustaining rural livelihoods shows in the assessments conducted (by MINAG) in 2006 and 2007 (and sponsored by Yanacocha), which both refer to the lack of water security to explain low agricultural productivity (INRENA, 2007).

When the first incidents of conflict happened in 2005, with Combayo opposing the expansion plans of the company, Yanacocha was prompt to initiate negotiations. Yanacocha's quick success in reaching an agreement with the mayor of Combayo "for joint work towards development and the protection of water quality and quantity in Combayo" largely happened because many in Combayo hoped the agreement would lead to improvements in water availability and security.[20] The agreement consisted of promises to: (1) support the implementation of a drinking water supply system for Combayo; (2) preserve the water sources of Combayo; and (3) develop social investment projects in the town.[21] This was the first agreement signed by the company. As noted, many comuneros were of the opinion that it was not respected, and this is what prompted the conflict of 2006.

The mediation process that followed the conflictive events of 2006 again ended with an agreement in which the state, in coordination with Yanacocha, promised to implement drinking water systems for the town of Combayo and its communities. In addition, the agreement stipulated that the prime minister would commission a water management study to be implemented at the river basin level. The idea was to assess and determine the water quality and quantity in the area and propose the best and most efficient ways to protect water resources and ensure water supply for Combayo for drinking and irrigation purposes. To develop these studies, the government engaged funds from the Inter-American Development Bank. The international consulting company Nippon Koei was hired to carry out these water studies in the Chonta and Mashcon Watersheds within 10 months. The consultants proposed several actions to manage water at the watershed level. These included the construction of a main reservoir (42.5 Hm^3) on the Chonta River to secure water for Cajamarca and irrigation for downstream areas of the watershed. They also proposed the construction of two minor reservoirs (about 1.5 Hm) at the upper side of Combayo to secure water for the Azufre Watershed.[22] Other proposed interventions in this regulated system included maintenance work on the irrigation canals and installation of water measurement devices (Koei, 2010).

The study was finished in 2010. It is now available online on the national water authority's website. Although they reduced the tensions, neither the agreement nor the post-conflict water study proposed interventions that would guarantee or improve the longer-term sustainability of water resources. They also had little resonance in the conflict area, as the agreement did little to influence what happened with the three lakes, nor did it propose solutions to the problems of water depletion in Combayo. Instead, the improvements seemed to depend on each community's political agency, its networks and its lobbying skills, as well as their success in mobilizing external funding from the company or other sources.

After the signing of the agreement, and during the preparations for the studies, Yanacocha proceeded with their actions in the Carachugo site. This entailed the draining of areas at the Ocucha Machay and Chaquicocha Creeks of about 412 ha and 685 ha, respectively. In May 2008, and in spite of the ongoing studies, the company also communicated to the water authority, ATDRC, that it would begin the removal of the three lakes. The ATDRC approved those actions and specified that the company would have to mitigate the reductions in water availability by releasing treated water, suitable for irrigation and animal consumption, to the creeks. To mitigate impacts at the Ocucha Machay Creek, the company would have to release water permanently (minimum discharges of 5 L/s in 2007, 15 L/s in 2009 and 35 L/s from 2011 onwards). Compliance with these agreements was to be controlled by the ATDRC.

On paper, these agreements sound reasonable, even though they do not include any concern about the longer-term sustainability of the water-based ecosystems. Yet, the capacity of the ATDRC to actually monitor and enforce them is highly doubtful. In fact, its little involvement during the Combayo conflict, its poor track record in managing water resources in the area (Caballero Martin, 2012), and its lack of credibility and legitimacy among affected communities (Sosa & Zwarteveen, 2012) seriously call into question whether the ATDRC will be able to make Yanacocha keep its part of the agreement or hold the company accountable for the impacts of its operations (Sosa & Zwarteveen, 2014). Some have suggested that the monitoring could also have been done by the technical committees established in 2000 by the ATDRC and appointed to assess and monitor water in areas where Yanacocha operates. However, because the technical committees are financially sponsored by the company, they have little credibility in the eyes of the rural population (Orian, 2008).

The quality of the water assessment studies themselves is likewise the subject of serious doubts and questions (Orian, 2008). As part of Yanacocha's expansion plans for the Carachugo site, the company carried out a water assessment study. This study characterized the three lakes that would be affected as intermittent (or non-permanent), reducing their significance. The study indicated that Patos Lake had a water volume that varied from 5135 m^3 during the rainy season to 2868 m^3 during the dry season, and that the other two lakes were basically empty during the dry season.[23] The deputy governor of Combayo considered these findings a strategic way to dismiss the relevance of the three lakes; it allowed the mining company to convince some leaders of Combayo that there was no water in the lakes. He stated apprehensively, The authorities supported the mine's version in exchange for money, and there is going to be a leach pad in place of the lakes!" The deputy governor was also disappointed and suspicious about the role of the ATDRC, because the water inventories elaborated and updated by this authority in 2007 did not mention that the three lakes were not registered. This omission, according to him, made it easier to forget about these waters.[24]

The representatives of the mining company countered the worries of the deputy governor about water availability in Combayo and the performance of Yanacocha by emphasizing the fact that there are water studies being carried out in the area by a renowned international consulting company and financed by the Inter-American Development Bank.

Conclusions

Although some *Combayinos* received benefits from the company, like temporary work or assignments for their communal companies, many things regarding local development in Combayo remained as they had been before the conflicts. In particular, nothing happened to better protect the water resources of Combayo. In spite of the promises and agreements, the mining company's operations depleted the three lakes that were at the centre of the conflict.[25] In the process, the communities lost faith in their collective ability to alter the course of mining events. Instead of the risky strategy to collectively mobilize to protect their water resources, they had come to appreciate that directly and individually dealing with the mining company to secure funding for their water projects would be more effective, at least in the short term. The conflicts indeed seemed resolved, but the underlying problems of environmental integrity and livelihood security are not.

To deal with the conflicts surrounding the activities in the highlands of Combayo, the government of Peru and Yanacocha importantly relied on two strategies. First, they made sure that what the company proposed to do was legally right just by obtaining all required permissions. This enabled the company to respond to complaints by simply stating that it was operating within the law. Second, Yanacocha made sure that what it did was scientifically sound, by conducting scientific impact assessments and proposing technical strategies to mitigate the impacts on the environment, particularly water. Yanacocha proposed for instance to compensate for depleted water sources by installing water treatment plants and by releasing treated water to communities suffering from the depletion. Together, these two strategies lent legitimacy to the company's operations by making them seem morally and scientifically sound. Indeed, the discussed examples show that legal compliance (justice) and technical (or scientific) accuracy function to legitimize mining operations, allowing the company to proceed with business as usual without having to take seriously the demands of ecosystems or communities. Rendering water problems legal and technical thus conveniently transforms them into problems that can be solved. It simultaneously renders them non-political (Li, 2007).

How this is problematic can perhaps best be illustrated with the example of the water assessment study that was proposed by the state as part of its conflict resolution strategy. Interestingly, the final report explicitly mentioned that its outcomes were dependent on how the problems were framed and by whom. According to the report, the fact that there were different parties involved, with diverging views and opinions, made it difficult if not impossible for the report to meet everyone's expectations.

> Since the launch meeting in June 2008, it became apparent that there were different and conflicting expectations among the actors [and] about the importance and the value of the study. As stated in public meetings, the city of Cajamarca and the water users [of the Chonta Watershed] expect the report to justify the need for a large dam in the Chonta Watershed to provide water not only for the city of Cajamarca but also to extend the irrigated areas close to the city. Exactly the opposite idea was echoed by the highland water users [Combayo and those nearby the mining operations], who hope that the study will emphasize the need for numerous reservoirs in the upstream areas of the watershed [to secure water for them]. [This action, however,] would have a direct negative effect on the amount of water that reaches the downstream areas. (Nippon Koei, 2010, p. 142)

The consultants thus acknowledged that, although they had made efforts to make the study as participatory as possible,[26] the fact that different parties had widely diverging and sometimes opposing views made it difficult to fully involve them and their interests: "Although an important mobilization of public opinion and information have been generated, a [comprehensive] response of actors' proposals has not been achieved, [nor have there been] organizational actions that could allow to work with a [legitimate] representative in the study area" (p. 142).

The report clearly showed that problems were articulated differently by different actors. Its interviews revealed that many *comuneros* and the authorities representing them wondered whether "the study would produce more water for the users, particularly from the upstream areas" or be "just another study" done in the area (p. 134).[27] Hence, while the mining company could use the study as an objective statement of fact, the very consultants conducting it were aware of its partiality. They were worried about the effectiveness and value of their own report, because they were conscious of the impossibility of screening off their analysis from the political context in which it was conducted.

This example serves to underscore the more fundamental point we want to make about the success of water conflict resolution. We have argued that the complex and deeply political nature of mining conflicts makes it difficult or perhaps even impossible to institutionally or scientifically safeguard objectivity and neutrality. Our analysis shows that there is a danger that rather than correcting unsustainable or unjust behaviours, supposedly neutral or objective solutions work and are used as a legitimizing device for those in power (in this case the mining company), to continue their business as usual.

Hence, rather than relying on forms of 'objectification' (law, science, technical solutions) that deny (eliminate, erase or render invisible – Edmunds and Wollenberg 2001) the intrinsically political nature of conflicts, water conflict resolution strategies should be much more explicitly concerned with the question of how to democratically organize political decision making processes, including the question of how to organize possibilities of objecting. This goes far beyond public hearings and stakeholder engagement. It also requires thinking beyond quick solutions (Himley, 2014) or short-lived forms of consensus, both of which tend to blur the diversity of positions and mask abuses of power (Castro, 2007; Edmunds & Wollenberg, 2001; Moreyra & Wegerich, 2006). Rather than seeking to neutralize differences in position and power, our analysis suggests that the longer-term sustainability of livelihoods and ecosystems may be better served by openly accepting and dealing with such differences, and by learning to acknowledge that experiences and knowledge (including science) are always contextually embedded and plural.

One implication is that effective environmental conflict resolution and water governance strategies should pay more attention to processes and power dimensions in conscious attempts to create a more or less level playing field. As Budds (2014) suggests, communities' abilities to engage and object require not just improved access to information, but also improvements in their skills to critically analyze and understand this information, as well as the capacity and the influence to use it and make it count. Another implication of our analysis is that creative ways need to be identified to give voice to the environment (the ecosystem) beyond the single voice of science, allowing it to speak in multiple ways (as articulated by the different parties involved). And a third important implication is that it becomes essential to find innovative ways of accounting for water uses beyond mere economic or market benefits, challenging dominant approaches of dealing with water (Trottier & Brooks, 2013) to include longer-term and often harder-to-measure values and functions.

Notes

1. This hacienda was the property of Eloy Santolalla, well known in the area for his mining activities in other areas of Cajamarca (Santolalla, 1906). The Land Reform of 1969 affected and dissolved this hacienda, like many others in the region.
2. Because of the proximity to Yanacocha, many households in Combayo rely on employment with the company. Usually farmers are hired for short periods (three to six months) as unskilled workers. During the development of the mining operations in the region, and encouraged by the company, farmers also created small community or communal companies to provide services to Yanacocha.
3. Together with the Grande and Quinuario and Paccha Rivers, the Azufre River forms the Chonta Watershed. The Chonta River is about 39.8 km long and together with the Mashcon River feeds into the Cajamarquino River, one of the most important in the region (Koei, 2010).
4. The irrigation canals of this river basin belong to the Chonta Water User Association (Junta de Usuarios del Río Chonta, JURCH). The water user associations in Combayo, however, are not formally part of JURCH, because the fees asked by the JURCH (30–40 soles) are too high for them and farmers do not feel that JURCH works for their benefit (CEDEPAS, 2008; Mendoza Moreno, 2008).
5. Interview by Alicia Abanto Cabanillas (commissioner in the Cajamarca Ombudsman office), with Luciano Llanos (then mayor of Combayo), and the main leaders of Combayo, 2 August 2005.
6. Focus group discussion with farmers from Bellavista Baja, 5 April 2009.
7. This dam was constructed to prevent mine sediments from obstructing the canals and water flows, but the people thought it was going to be for securing and increasing water quantity in the area. Personal communication, farmer leader from Bellavista Baja, 30 April 2009.
8. For a discussion of Peruvian national police being hired by private mining companies, see Kamphuis (2012).
9. The Peruvian legislation considers kidnapping a complex crime, with a punishment of imprisonment for 20–30 years (Sala Penal Cajamarca, 2010).
10. Different representatives from communities, civil society organizations and the private sector, as well as the government, argued that the Combayo conflict was more about communities' attempts to get more economic benefits from the mining company than

     about water and the environment. Years after the conflict, a representative of Yanacocha commented: "The conflict in Combayo was extortion; the trigger was not water but employment" (personal communication, 21 December 2010).

11. This meeting brought together high-level authorities, such as the ministers of agriculture, energy and mining, health, and economy and finance; five representatives of the parliament; the president of the regional government; the mayor of Cajamarca; the mayor of the Encañada District; about 50 representatives or leaders of the *caseríos* of Combayo; and 4 representatives of Yanacocha (Málaga Málaga, 2006; Villar et al., 2006).

12. Personal communication, former Yanacocha worker, March 2012.

13. The EIAs are prepared for the mining companies by consultant companies visiting the areas to be mined. During the presentations of the EIAs, these consultants inform the population about the activities to be developed by the company during operations.

14. During fieldwork and together with representatives of the Ombudsman office, M. Sosa attended a public hearing on an EIA for large mining exploration activities in Cajamarca. The representatives commented that perhaps the technical language used to explain environmental and water issues during the hearing made it difficult for many of those present to understand what was explained. Because income issues were more tangible and easier to grasp, some of them seemed less interested in environmental and water issues and instead started talking about issues of employment generation. Of the questions posed by the public, 9 were about water issues and more than 20 related to employment. For a discussion of the limitations of these public hearings as participatory events, see Li (2009) and De Echave et al. (2009).

15. Administrative resolutions 051-2008-INRENA-IRH and 367-2008-INRENA-IRH.

16. Deputy governor of Combayo, letters N.009 and N.010, CPM-Combayo, May 2008.

17. He commented that "precisely the days that the personnel of Yanacocha was drying up the lakes in the headwaters area ... the mine organized together with our mayor ... a music parade which included folkloric artists". Presumably his intention with this comment is to notice that water issues were blurred by other activities that were organized in Combayo at the same time.

18. Personal communication, 30 April 2009.

19. Personal communication with one of the judicially denounced farmers, 21 March 2009.

20. Letter N.009, CPM-Combayo, May 2008.

21. Yanacocha offered USD 1,500,000 to invest in Combayo, and the municipality arranged lists of people and companies from Combayo to work for Yanacocha (agreement documents from 15 and 20 September 2005).

22. In other part of the study, however, the proposal of constructing upstream reservoirs was problematic because this would affect water availability for downstream areas. The proposal to develop minor reservoirs is not new. Previous studies developed by the Water Authority in Combayo already proposed that alternative, but with differences concerning the selection of water sources.

23. Communication from Yanacocha to the ATDRC requesting approval for hydrological and hydrogeological studies of the Patos and Estación 1 lakes and approval for a draining plan, 11 March 2008.

24. Personal communication, 22 March 2009.

25. Yanacocha releases treated water to the creeks and the Azufre River, as stated in the permissions given by the state.

26. The document states that several information meetings and water quality monitoring were done with the participation of Combayo's population as well as public and private organizations of Cajamarca.

27. Since the 2000s, several water assessments have been done in Cajamarca (and in Combayo) by national authorities (INRENA, 2007) as well as by international bodies. For example, the Office of the Compliance Advisor/Ombudsman of the International Finance Corporation (CAO, 2007) commissioned a water study in the area by Stratus Consulting in 2003.

Acknowledgements

We thank all our interviewees in Cajamarca and Combayo for their predisposition to share their stories during the fieldwork. Thanks to Carlos Cerdán for his help with designing and drawing the map to illustrate this article. We thank also the editors of the special issue and the anonymous reviewer for their thoughtful comments and suggestions to improve the article.

Funding

This research was funded by the Netherlands Organisation for Scientific Research (NWO) [grant number W 01.65.308.00].

References

Arana, M. (2009). *Conflictos mineros, responsabilidad social empresarial e institucionalidad ambiental en Perú*. Paper presented at the Replanteando la industria extractiva: Regulación, despojo y reclamos emergentes, York University, Canada

Bebbington, A. (Ed.). (2007). *Minería, movimientos sociales y respuestas campesinas: una ecología política de transformaciones territoriales*. Lima: IEP CEPES, Centro Peruano de Estudios Sociales.

Bebbington, A., Hinojosa, L., Humphreys Bebbington, D., Burneo, M. L., & Warnaars, X. (2008). Contention and ambiguity: Mining and the possibilities of development. *Development and Change, 39*(6), 887–914. doi:10.1111/j.1467-7660.2008.00517.x

Bebbington, A., Humphreys Bebbington, D., & Bury, J. (2010). Federating and defending: Water, territory and extraction in the Andes. In R. Boelens, D. Getches, & A. Guevara (Eds.), *Out of the mainstream: The politics of water rights and identity in the Andes* (pp. 307–327). London: Earthscan.

Bebbington, A., & Williams, M. (2008). Water and mining conflicts in Peru. *Mountain Research and Development, 28*(3/4), 190–195. doi:10.1659/mrd.1039

Boelens, R. A. (2008). *The rules of the game and the game of the rules: normalization and resistance in Andean water control* (PhD Thesis). Wageningen University, Wageningen.

Bridge, G., & Perreault, T. (2009). Environmental governance. In N. Castree, D. Demeritt, D. Liverman, & B. Rhoads (Eds.), *A companion to environmental geography* (pp. 475–497). Oxford, UK: Wiley-Blackwell.

Budds, J. (2014). Acceso al agua y justicia hídrica: un análisis de las relaciones de poder entre Southern Copper Corporation y comunidades rurales en Moquegua y Tacna, Perú. In T. Perreault (Ed.), *Minería, Agua y Justicia Social en los Andes: Experiencias Comparativas de Perú y Bolivia* (pp. 41–58). Cusco: Centro Bartolomé de las Casas.

Burawoy, M. (1998). The extended case method. *Sociological Theory, 16*(1), 4–33. doi:10.1111/0735-2751.00040

Bury, J. (2002). Livelihoods, mining and peasants protests in the Peruvian Andes. *Latin American Geography, 1*(1), 3–19.

Bury, J. (2004). Livelihoods in transition: Transnational gold mining operations and local change in Cajamarca, Peru. *The Geographical Journal, 170*(1), 78–91. doi:10.1111/j.0016-7398.2004.05042.x

Caballero Martin, V. (2012). *Conflictividad social y gobernabilidad en el Perú*. La Paz: Programa de las Naciones Unidas para el Desarrollo, & Instituto Internacional para la Democracia y la Asistencia Electoral.

CAO. (2007). Building consensus: History and Lessons from the Mesa de Diálogo y Consenso CAO-Cajamarca, Peru. Monograph 2. The independent water study (2002–2004). Office of the Compliance Advisor/Ombudsman CAO - IFC.

Castro, J. E. (2007). Water governance in the twentieth-first century. *Ambiente & Sociedade, X* (2), 97–118.

CEDEPAS. (2008). *Diagnóstico socio económico de la cuenca del río Chonta*. Preparado para la Junta de usuarios del río Chonta y Cajamarquino, CEDEPAS Norte, Cajamarca.

De Echave, J., Diez, A., Huber, L., Revesz, B., Lanata, X. R., & Tanaka, M. (2009). *Minería y conflicto social*. Lima: IEP, CIPCA, CBC, & CIES.

Deza, N. (2008). *Impactos socio economicos de la minería aurífera por lixiviación de pilas a tajo ierto en Cajamarca, 1992–2007* (Tesis Doctoral en Ciencias Ambientales). Universidad Nacional de Trujillo, Trujillo.

Edmunds, D., & Wollenberg, E. (2001). A strategic approach to multistakeholder negotiations. *Development and Change, 32*(2), 231–253. doi:10.1111/1467-7660.00204

Guardia Nogales, A. (2011). *Exploring the veins of development: Politics of large-scale transnational mining, local access to water and small-scale agriculture in Cajamarca-Peru* (Master thesis). CEDLA, University of Amsterdam - Concertacion, Wageningen University.

Himley, M. (2014). Los límites de la solución tecnológica: Minería, agua y poder en el Perú. In T. Perreault (Ed.), *Minería, Agua y Justicia Social en los Andes: Experiencias Comparativas de Perú y Bolivia* (pp. 50–79). Cusco: Centro Bartolomé de Las Casas.

INRENA. (2007). *Estudio de priorización y selección de alternativas de embalse en la cuenca del Río Azufre, Combayo - Cajamarca* (Vol. I - Informe Principal). Lima: Ministerio de Agricultura- Instituto de Recursos Naturales (INRENA). Intendencia de Recursos Hídricos. Oficina de Proyectos de Afianzamiento Hídrico.

Kamphuis, C. (2012). Foreign mining, law and the privatization of property: A case study from Peru. *Journal of Human Rights and the Environment, 3*(2), 217–253. doi:10.4337/jhre.2012.03.03

Li, F. (2009). Documenting accountability: Environmental impact assessment in a Peruvian mining project. *PoLAR: Political and Legal Anthropology Review, 32*(2), 218–236. doi:10.1111/j.1555-2934.2009.01042.x

Li, T. M. (2007). *The will to improve: Governmentality, development, and the practice of politics*. Durham, NC: Duke University Press.

Lingán, J. (2008). El Caso de Cajamarca. In M. Scurrah (Ed.), *Defendiendo derechos y promoviendo cambios: El Estado, las Empresas Extractivas y las Comunidades Locales en el Perú*. Lima: Oxfam International, Instituto del Bien Común, & IEP.

Málaga Málaga, F. (2006). Atención Comisión Alto Nivel Caso Combayo. Dirección General de Gestión Social - Ministerio de Energía y Minas.

Manson, J. (2002). *Qualitative researching*. London: SAGE.

Mendoza Moreno, I. A. (2008). *Gobernabilidad del agua y actividad minera en la microcuenca del río Azufre. Cajamarca 2005–2007* (PhD Thesis). Universidad Nacional de Cajamarca, Cajamarca.

Mollinga, P. (2008). Water, politics and development: Framing a political sociology of water resources management. *Water Alternatives, 1*(1), 7–23.

Moreyra, A., & Wegerich, K. (2006). Highlighting the 'Multiple' in MSPs: The Case of Cerro Chapelco, Patagonia, Argentina. *International Journal of Water Resources Development, 22*(4), 629–642. doi:10.1080/07900620600779657

Muradian, R., Martinez-Alier, J., & Correa, H. (2003). International capital versus local population: The environmental conflict of the Tambogrande mining project, Peru. *Society & Natural Resources, 16*(9), 775–792. doi:10.1080/08941920309166

MWH. (2012). III Modificación del EIA Ampliación del Proyecto Carachugo Suplementario Yanacocha Este. Lima.

Newmont. (2012). Annual Report 2012. Retrieved from http://s1.q4cdn.com/259923520/files/doc_financials/annual/2012-Annual-Report_v001_e75k8c.pdf

Nippon, K. (2010). Plan de gestión de los recursos hídricos en las cuencas Mashcón y Chonta con énfasis en el afianzamiento hídrico de las subcuencas Azufre, Paccha y Río Grande de Chonta, Cajamarca: Nippon Koei

Orian, E. (2008). *The Transfer of Environmental Technology as a Tool for Empowering Communities in Conflict; the case of Participatory Water Monitoring in Cajamarca, Peru* (Master Thesis). University of Manchester, Manchester, UK.

Panfichi, A., & Coronel, O. (2010). Conflictos Hídricos en el Perú 2006–2010: Una lectura panorámica. In R. Boelens, L. Cremers, & M. Zwarteveen (Eds.), *Justicia Hídrica*. Lima: IEP.

Pascó-Font, A., Diez Hurtado, A., Damonte, G., Fort, R., & Salas, G. (2003). Aprendiendo mientras se trabaja. In G. McMahon & F. Remy (Eds.), *Grandes minas y la comunidad: efectos socio económicos en Latinoamérica, España y Canadá* (pp. 145–201). Bogota: Banco Mundial, IDRC-CRDI, & Alfaomega.

Peluso, N. L., & Watts, M. (2001). *Violent environments*. Ithaca: Cornell University Press.

Pereyra Matsumoto, C. (2008). Conflictos regionales e intersectoriales por el agua en el Perú. In A. Guevara Gil (Ed.), *Derechos y conflictos de agua en el Perú* (pp. 81–99). Lima: Concertacion, Walir, & Departamento Académico de Derecho - PUCP.

Ranjit, K. (2011). *Research methodology. A step-by-step guide for beginners* (3er Edition ed.). Los Angeles: SAGE.

Sala Penal Cajamarca. (2010). Caso Combayo. Corte de Justicia de Cajamarca. Cajamarca.

Santolalla, F. M. (1906). *Departamento de Cajamarca: Monografía geográfico-estadística*. Cajamarca: Impr. y Librería de San Pedro.

Sosa Landeo, M. (2012). La influencia de la gran minería en Cajamarca y Apurímac, Perú: Acumulación por despojo y conflictos por el agua. In E. Isch López, R. Boelens, & F. Peña (Eds.), *Agua, injusticia y conflictos*. Lima: IEP.

Sosa, M., & Zwarteveen, M. (2012). Exploring the politics of water grabbing: The case of large mining operations in the Peruvian Andes. *Water Alternatives, 5*(2), 360–375.

Sosa, M., & Zwarteveen, M. (2014). The institutional regulation of the sustainability of water resources within mining contexts: Accountability and plurality. [SI: Sustainability science - Legal pluralism]. *Current Opinion in Environmental Sustainability, 11*, 19–25. http://dx.doi.org/10.1016/j.cosust.2014.09.013

Tanaka, M., & Meléndez, C. (2009). Yanacocha y los reiterados desencuentros: Gran afectación, débiles capacidades de acción colectiva. In J. De Echave, A. Diez, L. Huber, B. Revesz, X. R. Lanata, & M. Tanaka (Eds.), *Minería y conflicto social*. Lima: IEP, CIPCA, CBC, & CIES.

Torres, V. (2007). *Minería artesanal y a gran escala en el Perú: El caso del oro*. Lima: CooperAcción - Acción solidaria para el desarrollo.

Trottier, J., & Brooks, D. B. (2013). Academic tribes and transboundary water management: Water in the Israeli-Palestinian peace process. *Science & Diplomacy, 2*(2), 1–13.

Urteaga, P. (Ed.). (2011). *Agua e industrias extractivas: Cambios y continuidades en los Andes*. Lima: IEP, & Concertación.

Villar, A., Gonzales, A., & Roncal, A. (2006). *Vigilancia Ciudadana a la Conflictividad y Conflictos en el marco de la presencia de Minera Yanacocha en Cajamarca - "Caso Combayo"* (Vol. 3). Cajamarca: Cedepas Norte, Propuesta Ciudadana, & Oxfam America.

Yanacocha. (2007). *Yanacocha balance social y ambiental - 2006*. Cajamarca: Minera Yanacocha S.R.L.

Yanacocha. (2008). *Cajamarca, tierra fecunda. Balance social y ambiental - 2007*. Cajamarca: Minera Yanacocha S.R.L.

Zavaleta, M. (2014). La batalla por los recursos en Cajamarca. Cuaderno de Trabajo N. 18. Lima: Departamento de Ciencias Sociales - PUCP.

Zwarteveen, M. (2015). *Regulating water, ordering society. Practices and politics of water governance*. Inaugural lecture, University of Amsterdam, Amsterdam.

Zwarteveen, M., Boelens, R., & Roth, D. (2005). Anomalous water rights and the politics of normalization. Collective control and privatization policies in the Andean region. In D. Roth, R. Boelens, & M. Zwarteveen (Eds.), *Liquid relations. Contested water rights and legal complexity*. New Brunswick: Rutgers University Press.

Zwarteveen, M. Z., & Boelens, R. (2014). Defining, researching and struggling for water justice: Some conceptual building blocks for research and action. *Water International, 39*(2), 143–158. doi:10.1080/02508060.2014.891168

Predicting water quality associated with land cover change in the Grootdraai Dam catchment, South Africa

Anja du Plessis, Tertius Harmse and Fethi Ahmed

The Grootdraai Dam catchment forms part of the Vaal River system, which is deemed to be the 'workhorse' of South Africa as it is located within the economic heart of the country. The status of water quality within the catchment is an important characteristic that needs to be investigated extensively due to its importance to the country's future economic growth. Intricate relationships between land cover and specific water quality parameters were quantified and unique model equations were formulated to predict water quality in the region. Urban and mining developments should be re-evaluated due to the accompanied significant hydrological consequences.

Introduction

The quality of the world's water resources and the accompanied freshwater and coastal ecosystems have been negatively affected by primarily anthropogenic factors due to increased nutrient loading and, in some cases, pathogens within these affected water bodies, to name two. Changes in land cover have consequently led to the loss of biodiversity through the modification, fragmentation or loss of habitats, the degradation of the soil and water resources, and the overexploitation of endemic or native species (Deelstra, Oygarden, Blankenberg, & Eggestad, 2011; Meador & Goldstein, 2003; Rothenberger, Burkholder, & Brownie, 2009; Seeboonruang, 2012). Population pressures and intensified economic activity in sub-basins have altered land-use patterns and generated water pollution (Jung, Lee, Hwang, & Jang, 2008; Tong & Chen, 2002; Wilson & Weng, 2010).

The relationship between land use and water quality within different regions or catchments has been widely researched across the world (Jung et al., 2008; Wilson & Weng, 2010). Extensive research has been completed on the relationships between land cover or land-use change and different water quality parameters. Consequently, studies have progressed from identifying relationships to predicting water quality in relation to land cover change or determining and comparing the accuracy or uncertainty of various river water quality modelling techniques such as the partial least squares (PLS) regression model (Baker, 2011; Hall, Germain, Tyrrell, & Sampson, 2008; Huang & Klemas, 2012; Li, Gu, Lui, Han, & Zhang, 2008).

South Africa is a semi-arid country with inefficient management of water that results in overall deterioration in its quality. This persists even though water is described as a vital resource in maintaining life and sustaining agriculture, manufacturing, transportation and other economic activities (Ashton & Haasbroek, 2002; DEAT, 2005; Helmschrot & Flugel, 2002; Kadewa, Moyo, Mumba, & Phiri, 2005; King, Maree, & Muir, 2009; Usali & Ismail, 2010).

The current and anticipated growth rates of the population and trends in future socio-economic development indicate that South Africa's freshwater resources will be unable to sustain the current patterns of water use and discharge. This has consequently led to numerous water supply problems across the whole country. At present, multiple water management areas are experiencing a water deficit. The natural ecosystems as well as the region's and the country's freshwater resources are put under immense pressure by various sectors and users, which lead to the overall deterioration of water resources (Ashton, Hardwick, & Breen, 2008).

This study has consequently focused upon the quantification of these relationships between selected water quality parameters and land cover change with the application of a partial least squares (PLS) regression to formulate model equations for the prediction of water quality and possibly assist with the development of a presently lacking science–policy interface. This study, therefore, puts forward a practical application of these formulated model equations for the prediction of water quality within a specific catchment. It presents the importance of the formulation of the PLS model equations to promote informed and accurate decision-making within the water management sector. It also presents a practical application of these within a catchment that has been plagued by immense land cover change due to mining and urban developments accompanied with poor and uninformed decision-making processes.

Study area

The Grootdraai Dam catchment, which is the main focus of this research paper, is situated within the Upper Vaal Water Management Area and located at the upper reaches of the Vaal River, which is considered to be the main water source for the central industrial, mining and metropolitan regions in South Africa (Figure 1). The Grootdraai Dam catchment together with the rest of the Vaal River catchment serves six of the nine provinces of the country, and is consequently of great importance in terms of the environmental, social and economic spheres (Bertasso, 2004; DWAF, 2004). The region is presently plagued by numerous applications for prospected mining rights and urban sprawl, which may be accompanied by immense degradation of the region's freshwater resources.

The natural landscape of the Grootdraai Dam catchment has therefore been transformed and manipulated physically and chemically in order to meet society's needs. The changes of land cover have been accompanied by various impacts on a specific region's hydrological responses and ultimately its water resources. Increasing demands from society as well as economic pressures have been accompanied with further land cover changes within the region and have recently contributed to water availability and water quality problems within the catchment (Attua, Ayamga, & Pabi, 2014; Warburton, Schulze, & Jewitt, 2012).

Research done on the Grootdraai Dam catchment was found to be minimal. Most research has focused upon the identification of causes and consequences of water

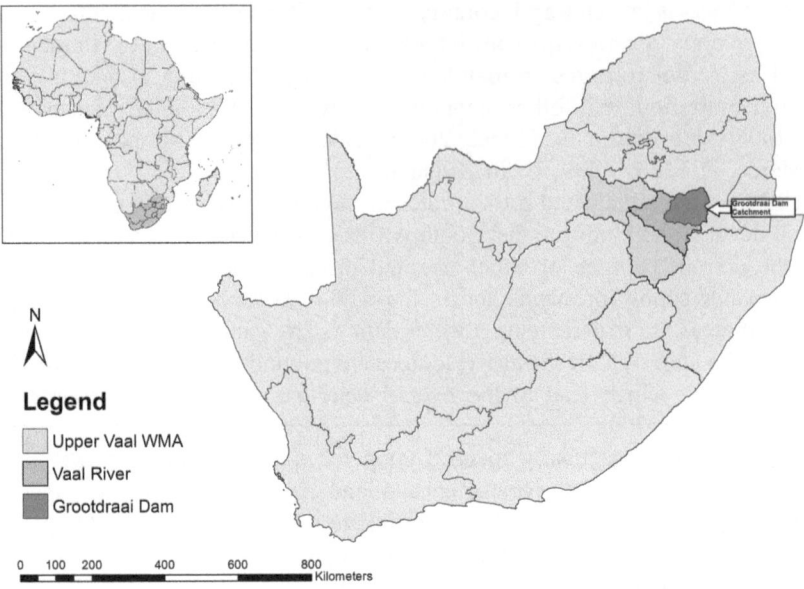

Figure 1.   Location of the Grootdraai Dam catchment within the Vaal River catchment, South Africa.

degradation within the catchment with a primary focus upon mining operations as well as on other water quality problems such as acid mine drainage (McCarthy, 2011; Van Steenderen, Theron, & Hassett, 1987).

The predominant methodologies implemented by previous research have been a combination of qualitative and quantitative methods. Qualitative methods, which include literature and observations, were aimed at describing the current state of the larger Vaal River catchment as well as the development of management plans to be implemented by authorities. Quantitative methods included basic descriptive statistics aimed at identifying possible pollution sources and areas of concern. This research therefore proposed that the interactions between hydrological responses, in terms of water quality, and land use needs to be quantified for the catchment or region to improve upon its water resource planning and management through the application of PLS regression analysis and formulated model equations. The formulated PLS regression model equations would therefore enable the prediction of the selected water quality parameters in relation to a change in land cover and can hopefully aid in the development of a science–policy interface.

This research investigated the hydrological responses, in terms of water quality, in the case of land cover change within the Grootdraai Dam catchment due to its ecological, social and economic importance, but also for its definite widespread negative cumulative impacts on the environment and human health. Relationships and unique model equations were established, formulated and applied in a possible future land cover change scenario to predict the future water quality of the catchment, which has not been completed for the catchment by any previous research. The results can ultimately be used to determine what sustainable measures need to be taken in terms of land cover change and identify possible policy changes that need to be taken to promote a more sustainable Grootdraai Dam catchment and ultimately the larger

Vaal River catchment through improving informed decision-making and management processes.

Methods

The research made use of rainfall, evaporation, water flow and water quality data for the period July 2000–June 2012 as well as land cover data for 1994, 2000, 2005 and 2009 to establish the hydrological responses in terms of water quality in the case of land cover change.

Rainfall, water flow and evaporation data

The rainfall data were obtained from the South African Weather Service for the period July 2000–June 2012. Daily readings were used to calculate the average for the month. The average monthly rainfall figures were obtained for each of the two weather stations. Sample stations with long periods of no recorded data were excluded to ensure good-quality data and sound results. Water flow, as well as evaporation data, were obtained from the Department of Water Affairs (DWA) for July 2000–June 2012. The daily average water flow rate (m^3/s) was measured at all the river water flow sample stations over this period. Monthly water flow and evaporation values were calculated for each of the stations. Sample stations with large data inconsistencies, as well as long periods of no recorded data, were excluded to ensure good quality and accurate data.

Water quality data

Water quality data were obtained from all the available Rand Water (water services provider for the Upper Vaal WMA) at sample points situated in the Grootdraai Dam catchment for the same period. A total of 12 water-quality sample stations were used and provided a complete dataset of water quality data for this research. The locations of these are set out in Figure 2.

The water quality data obtained from these sampling points were measured monthly, weekly and, in some cases, daily throughout the year, but at no scheduled time and on no fixed day or week. A monthly average was calculated for each of the water quality parameters used at each station.

Sampling stations with inadequate data recordings were excluded. They were excluded in the case of the station having recorded fewer than four measurements within a year. Some water quality sampling stations were also not included in this research due the station having measured fewer than four parameters with fewer than four recorded measurements within a year. The 'Four by Four' (4 × 4) rule was therefore followed. It ensures that only sampling stations that regularly monitor the relevant parameter are included and eliminates stations with only three monitoring phases per year. This research was therefore limited to these sampling stations that had adequate replication in an attempt to ensure high-quality data and representation within the Grootdraai Dam catchment.

A wide variety of water quality parameters was used in order to obtain a holistic and accurate view of a water body's water quality in terms of environmental and human health. Water quality parameters to be used in this research were selected according to a selection process and were selected according to the following rules:

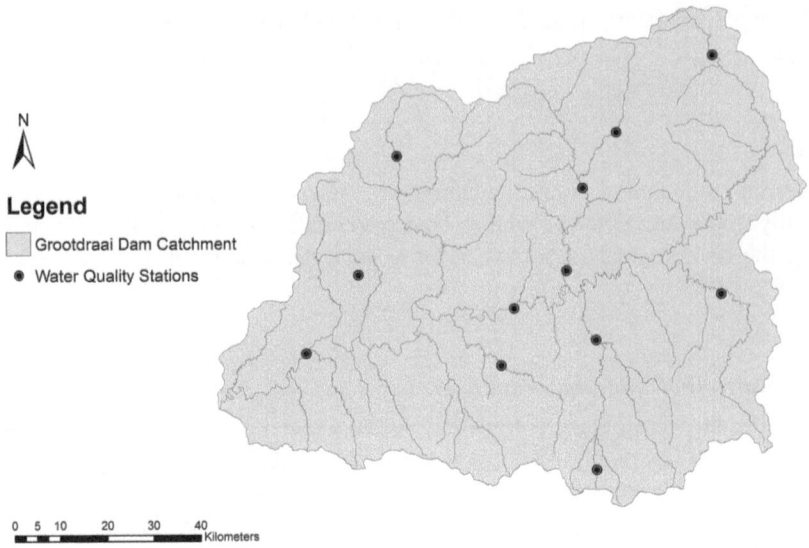

Figure 2. All the available Rand Water quality sampling stations used within the Grootdraai Dam catchment.

- The water quality parameter needs to have available national or regional water quality indices or guidelines.
- The water quality parameter must have been commonly measured and reported by Rand Water's water quality sampling stations within the Grootdraai Dam catchment.
- The water quality parameter needs to have a representation percentage of a minimum of 100%, 50% in the case of biological parameters, within the Grootdraai Dam catchment.
- Water quality parameters characterized by the occurrence or measurement of non-detectable values needs to be excluded due to the possibility of bias.

This research made use of relevant national and regional water quality standards and guidelines. The Vaal Dam in-stream water quality guidelines were predominantly used. In the event of a water quality parameter not having a regional standard, the appropriate South African national water quality standard was used and indicated accordingly. Water quality parameters were therefore excluded on the basis of the availability of international, national and regional guidelines. Subsequently, a total of 14 water quality parameters were identified, according to the described rules, for this research and include the following:

- *Physical* parameters: pH and electrical conductivity (EC).
- *Chemical* parameters: alkalinity, calcium, chloride, sodium, magnesium, nitrate, sulphate, ammonia, phosphate and chemical oxygen demand (COD).
- *Biological* parameters: dissolved oxygen (DO) and dissolved organic carbon (DOC).

The relevant national and regional water quality standards and guidelines were applied in the establishment of water quality status of the Grootdraai Dam catchment. The colour

Table 1.   Applied colour classification of water quality standard.

Water quality standard	Colour class
Ideal	
Acceptable	
Tolerable	
Unacceptable	

classification was used according to the relevant water quality standards and catchment guidelines, as shown in Table 1.

Land cover data

National land cover data were obtained from the Agricultural Research Center (ARC) for 1994, 2000, 2005, while land cover data for 2009 were obtained from the South African National Biodiversity Institute (SANBI) (ARC, 2005; SANBI, 2009). These land cover datasets were developed to establish the land cover for the whole of South Africa (no other datasets are available) and were consequently used in order to establish land cover changes for the Grootdraai Dam catchment. The land cover data from 1994 were included in the research to obtain a baseline land cover characteristic for the catchment. Land cover data from 2000, 2005 and 2009 were used in the PLS correlation and regression analysis.

The national land cover data used the Standard Land Cover Classification Scheme as the reference system to compile a 49-class legend. This full classification scheme is based on a hierarchical framework designed to suit the South African environment and incorporates known land cover types, which can be identified in a consistent and repetitive manner from high-resolution satellite imagery such as LandSat TM and SPOT (Thompson, 1996).

In their turn, the class definitions ensured that the data were standardized, and that broad generic classes were subdivided into more specific user-defined subclasses. The classification used by the land cover data was designed to conform to internationally accepted classification standards and conventions to ensure cross-border compatibility and integration with existing national and international land cover classification systems and datasets.

This research made use of the five-class legend used by the 2005 land cover dataset in order to describe the land cover. The seven-class legend for the land cover of 1994, 2000 and 2009 needed to be converted into the five-class legend used for 2005 during data analysis in order to obtain uniformity. The five land-cover classes were based on Level I and II classes and include the following:

- Urban build-up.
- Forestry and Plantations.
- Mining.
- Cultivated.
- Other (Natural, Degraded and Water bodies).

The five class-legend land cover was used to establish the land cover change over the period and ultimately to determine relationships between water quality and land cover.

Note that the 'other' land cover class constitutes natural, degraded and water bodies land cover classes. 'Other' land cover is predominantly constituted of natural land cover followed by water bodies and degraded land cover classes.

Shortcomings were also identified regarding the use of national land cover data supplied by ARC and SANBI. These included possible inaccuracies as well as low spatial resolution or cell size regarding the 2005 land cover dataset. The accuracy of the land cover data could have been decreased due to the project being completed by multiple stakeholders. Over- or under-estimations of land cover or the inaccurate identification of land cover could therefore have taken place (Burrough, 1990; Thompson, 1996). A logical error test was completed through the completion of a sensitivity analysis between the datasets to establish whether significant errors were present. A sensitivity test was completed by ARC between 1994, 2000 and 2005. Consequently, this research completed a sensitivity analysis regarding land cover classes between the 2005 and 2009 datasets. Very few errors were however identified through this testing process. Errors were corrected when necessary.

Mixed classes or classification structure is also a recognized limitation of land-use data. Mixed classes or classification structure were not a limitation in this study due to the use of a hierarchical classification design namely the Standard Land Cover Classification Scheme for Remote Sensing Applications in South Africa, as described previously (Lo & Yeung, 2002; Longley, Goodchild, Maguire, & Rhind, 2005; Thompson, 1996).

Data analyses

Temporal analyses and basic descriptive statistical analyses (data summaries) were completed on average monthly rainfall, water flow, evaporation and water quality data to identify significant variations and possible relationships between all the mentioned variables for July 2000–June 2012. Rainfall and evaporation data measured by the weather and evaporation sampling stations were correlated with water quality data from water quality stations located within close proximity thereof. Water flow data measured by the water flow sampling stations were correlated with water quality data measured by the water quality sampling stations located at the same location or downstream of the applicable water flow sampling station.

A temporal trend analysis was also completed between water quality and land cover for the periods 2000, 2005 and 2009 upon which the PLS model equations are built on. The independent variables were therefore the land cover classes and the dependent variables, and the selected water quality parameters. A multivariate statistical analysis in the form of PLS correlation and regression analysis was completed between the selected water quality parameters and the five class-legend land cover for 2000, 2005 and 2009. The PLS analysis is a proven effective variance-based structural equation model approach recommended for small sample sizes or in the case when the number of explanatory variables exceeds the number of observations and a high level of multicollinearity (Ibrahim & Wibowo, 2013a; Nasser & Wisenbaker, 2003). The use of PLS is recommended as it addresses multivariable problems such as the ill-effect of multiplex between variables (Ibrahim & Wibowo, 2013a, 2013b; Lou, Zhao, Chen, & Zhao, 2009; Singh, Jakubowski, Chidister, & Townsend, 2013).

The application of the PLS analysis enabled the prediction of water quality according to land cover changes with the development of a PLS regression model and associated algorithm. PLS regression models therefore facilitated the establishment of unique equations for each water quality parameter at each water quality sampling station.

Average concentrations were calculated for each selected water quality parameter in terms of the period over which the land cover dataset was established. The average annual concentration levels of the water parameters of the studied region were therefore calculated for 2000, 2005 and 2009 (includes full seasonal year, i.e., July 2000/ 2004/2008–June 2001/2005/2009), and paired with the land cover class percentages for those same time periods. This was completed for each water quality sampling station.

A PLS correlation and regression analysis was completed between the percentage of each specific land cover class and the calculated concentration level of each of the selected water quality parameters at each water quality station for 2000, 2005 and 2009 to establish the respective relationships between the land cover classes and the selected water quality parameters.

An overall average was calculated with the use of all the water quality sampling stations for each of the identified water quality parameters, for 2000, 2005 and 2009 to coincide with land cover data, to obtain a generalized water quality status as well as to formulate a generalized model equation for the Grootdraai Dam catchment as a whole. PLS regression model equations were established to quantify the overall hydrological responses for each selected water quality parameter to land cover change and to predict the concentration levels of these parameters in the event of such change. The root mean square error (RMSE), which is a standard statistical metric to measure model performance in some geoscience studies such as air quality and other geosciences, was used in this research to indicate the average model performance for each model equation as it assists in providing a holistic view of the error distribution (Chai & Draxler, 2014).

These model equations are consequently used to predict the specific concentration of all the identified water quality parameters according to the percentage land cover in the Grootdraai Dam catchment. A simulation was completed to quantify and predict concentrations of the selected water quality parameters in accordance with a possible land cover change scenario for 2015, 2020, 2030 and 2050.

It should be noted that these formulated model equations were based on only three periods (2000, 2005 and 2009) due to limited datasets. Furthermore, there could also be instances of multicollinearity in these model equations due to high correlation amongst the explanatory values themselves. These are therefore identified shortcomings that can be improved upon in future research with the completion of the 2014 land cover dataset which will enable the validation of these formulated model equations. Note that the adjusted R^2 value and the RMSE value needs to be taken into account for each individual model equation as the accuracy of these models vary and is not uniform.

Results and discussion

The water quality of the Grootdraai Dam catchment is of concern in light of the tolerable and unacceptable water quality standards. The quality of the water in this catchment has been degraded and its standard is of concern due to some of the water quality parameters proving to be either tolerable or unacceptable in terms of the Vaal Dam in-stream water quality guidelines.

Table 2.   Average monthly concentrations of the selected water quality parameters in the Grootdraai Dam catchment in terms of the Vaal Dam in-stream water quality guidelines and standards.

Water quality parameter	Grootdraai Dam
pH (pH units)	7.90
EC (mS/m)	42.66
Alkalinity (CaCO$_3$ mg/l)	136.56
Calcium (mg/l)[a]	26.18
Chloride (mg/l)	23.89
Sodium (mg/l)[a]	25.91
Magnesium (mg/l)[a]	16.87
Nitrate (mg/l)	0.79
Sulphate (mg/l)	39.07
Ammonia (mg/l)	1.63
Phosphate (mg/l)	0.88
COD (mg/l)	29.40
DO (mg/l)[b]	5.31
DOC (mg C/l)[a]	8.77

Notes: [a]According to national domestic-use water quality guidelines.
[b]According to national aquatic ecosystems water quality guidelines.

Water quality

The Grootdraai Dam catchment is characterized by tolerable standards of EC and DO as well as unacceptable standards of alkalinity, ammonia, phosphate and COD, as indicated in Table 2.

The water quality of the Grootdraai Dam catchment varies significantly across the region. On account of its very poor water quality across the range of some of the selected water quality parameters, this catchment poses significant challenges, with the hydrological conditions here generally being considered to be a matter of concern. The main areas of concern proved to be those regions located in close proximity to or downstream of urban built up areas (includes industrial areas), mining as well as cultivated land cover. These areas should therefore be seen as areas of major concern and should consequently be monitored closely for future changes to ensure that the quality of the water in this catchment would not be degraded even further.

Land cover

Firstly, rainfall, evaporation and water flow data were included in the research as these variables may have an influence on water quality. A Pearson correlation as well as a PLS correlation analysis were completed between these variables to determine the significance of their effects on water quality. The research found that the effects of rainfall, evaporation and water flow are not significant over the time period ($p < 0.05$; $n = 132$). The research subsequently concluded that land cover will have the most predominant and significant impact on water quality as it is the main determinant regarding the type and significance of the relationships and is consequently the main focus.

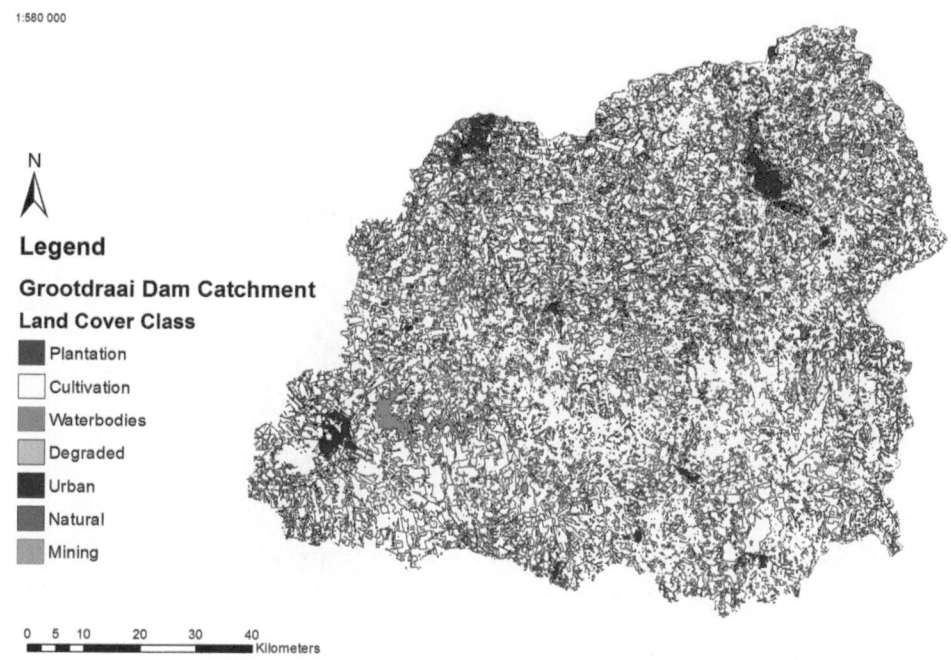

Figure 3.   Land cover within the Grootdraai Dam catchment, 2009.

As indicated in Figure 3, the Grootdraai Dam catchment is dominated by cultivated land. This land cover class constituted 82.97% of the total land cover in 2009. This was followed by the 'other' land cover category, which constituted 16.06% (10.49% of which was natural and 2.67% of which consisted of water bodies), mining land, which constituted 0.45%, and forests and plantations, which constituted less than 0.001% of the total land cover.

As shown in Figure 4, the catchment has undergone various land cover changes since 1994 and illustrates the catchment's dynamic nature. There was an upward trend in the urban built-up land cover class from 1994 to 2000, which has subsequently experienced a decline that may be due to the region having focused upon agricultural activities. The forestry and plantations land cover class has declined to such an extent that it no longer features in the catchment. In terms of mining, the catchment has generally experienced consistent growth, but also a decline. These trends will change in future, however, owing to the numerous prospected mining operations planned for this catchment. The cultivated and the 'other' land cover category experienced upward as well as downward trends. It should be noted that the immense decrease and increase of certain land cover classes such as cultivated and 'other' land cover from 2005 to 2009 were flagged as outliers and these values were adjusted accordingly to ensure accuracy within the analysis.

Quantifying hydrological change

It was established that the catchment is experiencing tolerable to unacceptable concentration levels of EC, COD, alkalinity, ammonia and phosphate. The recent temporal trends in the water quality within the catchment have shown an increase from 2009 in the concentration levels of EC, alkalinity, calcium, nitrate, sulphate and ammonia. The main possible agents for the raised concentrations of these water quality parameters were

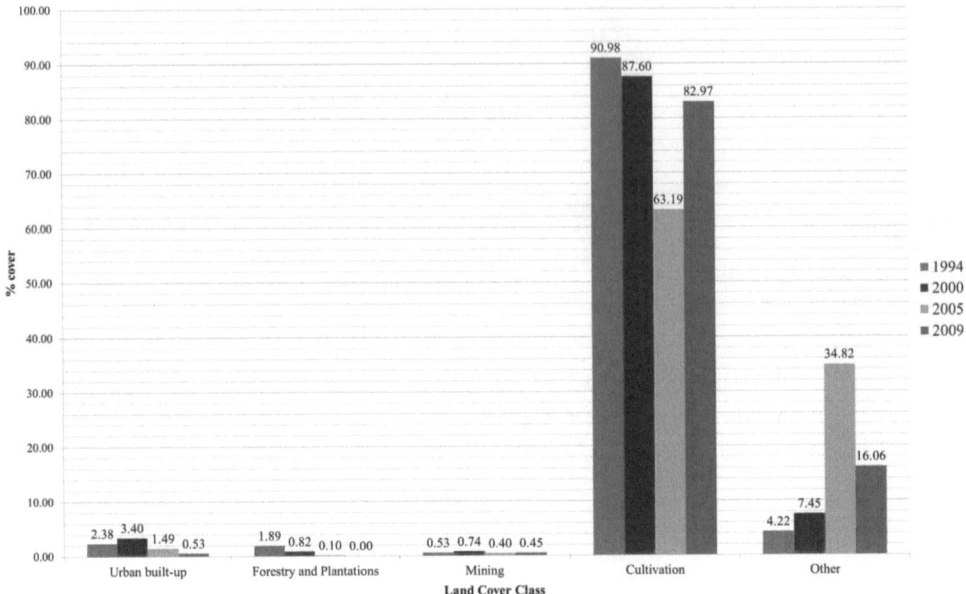

Figure 4.   Land cover trends within the Grootdraai Dam catchment, 1994–2009.

determined by establishing the relationships between the relevant variables, as indicated in Table 3. Values highlighted in **bold** are different from zero with a significance level alpha = 0.05, n = 3.

The very few significant relationships were found between water quality and land cover change within the Grootdraai Dam catchment. Negative relationships of

Table 3.   Partial least squares (PLS) correlation matrix obtained between land cover classes and the selected water quality parameters for the Grootdraai Dam catchment.

Variables	Urban built-up	Forestry and Plantations	Mining	Cultivation	Other
Urban built-up	**1.000**	0.975	0.895	0.360	−0.480
Forestry and Plantations	0.975	**1.000**	0.971	0.557	−0.662
Mining	0.895	0.971	**1.000**	0.739	−0.822
Cultivation	0.360	0.557	0.739	**1.000**	−0.991
Other	−0.480	−0.662	−0.822	**−0.991**	**1.000**
Ph	0.968	0.888	0.753	0.113	−0.244
EC	−0.396	−0.183	0.056	0.715	−0.616
Alkalinity	−0.815	−0.667	−0.470	0.247	−0.117
Calcium	**−0.998**	−0.986	−0.919	−0.413	0.530
Chloride	0.979	**1.000**	0.967	0.544	−0.650
Sodium	0.983	0.918	0.797	0.181	−0.310
Magnesium	−0.964	**−0.999**	−0.981	−0.594	0.695
Nitrate	0.711	0.849	0.950	0.912	−0.958
Sulphate	−0.568	−0.373	−0.140	0.564	−0.449
Ammonia	−0.629	−0.443	−0.216	0.499	−0.379
Phosphate	0.984	0.922	0.802	0.190	−0.319
COD	**1.000**	0.980	0.905	0.381	−0.500
DO	0.944	**0.994**	**0.992**	0.647	−0.742
DOC	0.714	0.542	0.326	−0.397	0.271

significance were found between calcium and urban built-up, as well as magnesium and forestry and plantations. Positive relationships of significance were established between COD and urban built-up, as well as between DO and forestry and plantations.

This research did however establish that an increase in the extent of the 'other' land cover category – issuing from the transformation of cultivated, urban built-up, as well as mining land cover classes – was found to be accompanied by undesirable and, in some cases, insidious effects. Furthermore, it was found that the buffering properties of the catchment tend to decline in tandem with the constant transformations (and re-transformations) that the land cover undergoes, especially in the case of the 'other' category. As a result, an increase in the extent of the 'other' land cover category within the catchment is expected to contribute to further degradation in the quality of the water in the catchment and needs to be highlighted.

The Grootdraai Dam catchment is currently showing a declining degree of resilience towards land cover change. Driving forces such as economic pressure, together with the further transformation of the 'other' land cover category into mining and urban built-up land cover, undesirable negative effects would be brought to bear on the surrounding water bodies.

Thus, the proclamation of numerous mining developments within the catchment is worthy of careful attention. In the event of the closure of a mine, it is of prime importance that the correct decommissioning measures be enforced to limit future challenges regarding the quality of the water within the catchment.

Model equations were established for each water quality parameter within this drainage region in terms of the land cover class proportions (percentages), as given in Table 4. These equations could be used, in association with the current trends in land cover change and in management practices, to predict the future concentration levels of these water quality parameters.

Predictions could therefore be made with a good amount of certainty and accuracy on future pH, calcium, chloride, magnesium, nitrate, ammonia, phosphate, COD and DO concentration levels within the Grootdraai Dam catchment due to the high adjusted R^2 values, indicating strong correlations between variables, and low RMSE values indicating low standard errors in these models. Model equations with a high adjusted R^2 value and a low RMSE value can therefore be used as a guideline when trying to predict changes in water quality due to land cover change and can also be incorporated in the development of a science–policy interface.

Future of the Grootdraai Dam catchment

The Grootdraai Dam catchment is experiencing an increase in the extent of its 'other' land cover category. This might develop into a situation that would be the direct opposite of the current conditions in the region – on account of the seemingly periodic nature of the relationship between the 'other' land cover category and the cultivated land cover class. As a result of future economic growth and development pressures, and of the approval granted by the numerous prospected mines, this catchment could experience an increase in urban built-up land, mining as well as cultivated land cover.

The following scenario was developed in terms of these assumed economic developmental trends and with the approval of the prospected mines in the catchment. It assumes a constant increase of 1% per year in the areal extent of the urban built-up land cover class, as well as in mining land cover. Table 5 presents the results for this scenario, which were obtained with the application of the model equations as given in Table 4.

Table 4. Model equations and the respective adjusted R^2 values and root mean square error (RMSE) values for each of the selected water quality parameters related to each land cover class for the Grootdraai Dam catchment.

Water quality parameter	Model equation	R^2	RMSE
pH	pH = 10.62 – 02 × Urban built-up + 9.75 – 02 × Forestry and Plantations + 0.22 × Mining + 1.28 – 03 × Cultivation – 1.57 – 03 × Other	0.61	0.099
EC	EC = 48.27 – 03 × Urban built-up + 1.86 – 02 × Forestry and Plantations + 4.18 – 02 × Mining + 2.44 – 04 × Cultivation – 2.99 – 04 × Other	0.00	1.000
Alkalinity	Alkalinity = 140.99 – 0.55 × Urban built-up – 1.76 × Forestry and Plantations – 3.95 × Mining – 2.31 – 02 × Cultivation + 2.82 – 02 × Other	0.26	1.000
Calcium	Calcium = 28.41 – 0.27 × Urban built-up – 0.86 × Forestry and Plantations – 1.93 × Mining – 1.13 – 02 × Cultivation + 1.38 – 02 × Other	0.87	0.415
Chloride	Chloride = 21.05 + 0.29 × Urban built-up + 0.93 × Forestry and Plantations + 2.10 × Mining + 1.23 – 02 × Cultivation – 1.50 – 02 × Other	0.96	0.258
Sodium	Sodium = 23.92 + 0.26 × Urban built-up + 0.84 × Forestry and Plantations + 1.89 × Mining + 1.10 – 02 × Cultivation – 1.35 – 02 × Other.	0.68	0.740
Magnesium	Magnesium = 17.70 – 0.11 × Urban built-up – 0.34 × Forestry and Plantations – 0.76 × Mining – 4.46 – 03 × Cultivation + 5.45 – 03 × Other	0.98	0.067
Nitrate	Nitrate = 3.03 – 02 × Urban built-up + 7.86 – 02 × Forestry and Plantations + 0.18 × Mining + 1.03 – 03 × Cultivation – 1.26 – 03 × Other	0.88	0.038
Sulphate	Sulphate = 40.98 – 0.146 × Urban built-up – 0.47 × Forestry and Plantations – 1.06 × Mining – 6.17 – 03 × Cultivation + 7.54 – 03 × Other	0.03	1.000
Ammonia	Ammonia = –0.34 – 02 × Urban built-up – 7.99 – 02 × Forestry and Plantations – 0.18 × Mining – 1.05 – 03 × Cultivation + 1.28 – 03 × Other	0.07	0.382
Phosphate	Phosphate = 3.39 – 02 × Urban built-up + 8.62 – 02 × Forestry and Plantations + 0.19 × Mining + 1.13 – 03 × Cultivation – 1.38 – 03 × Other	0.68	0.075
COD	COD = 28.04 + 0.20 × Urban built-up + 0.66 × Forestry and Plantations + 1.47 × Mining + 8.60 – 03 × Cultivation – 0.01 × Other	0.85	0.350
DO	DO = 13.83 – 03 × Urban built-up + 0.03 × Forestry and Plantations + 6.21 – 02 × Mining + 3.63 – 04 × Cultivation – 4.43 – 04 × Other	0.99	0.003
DOC	DOC = 8.81 + 4.66 – 03 × Urban built-up + 1.50 – 02 × Forestry and Plantations + 3.37 – 02 × Mining + 1.97 – 04 × Cultivation – 2.41 – 04 × Other	0.13	0.049

Table 5. Predictions of water quality parameter concentrations in the event of a 1%/year increase in urban built-up land cover within the Grootdraai Dam catchment showing water quality colour classification indicators in terms of the Vaal Dam in-stream water quality guidelines.

(a) Prediction of land cover (%)

Land cover class	2015	2020	2030	2050
Urban built-up	1.53	6.53	16.53	36.53
Forestry and Plantations	0.00	0.00	0.00	0.00
Mining	0.50	1.06	2.17	4.40
Cultivation	79.97	71.91	74.00	56.40
Other	18.00	20.50	7.30	2.67

(b) Prediction of concentration levels of water quality parameters

Water quality parameter	2015	2020	2030	2050	R^2	RMSE
pH (pH units)	7.82	7.99	8.41	9.16	0.61	0.099
EC (mS/m)	44.99	48.55	62.45	72.29	0.00	1.000
Alkalinity (CaCO3 mg/l)	145.55	153.21	171.35	204.24	0.26	1.000
Calcium (mg/l)[c]	31.70	32.65	34.90	38.97	0.87	0.415
Chloride (mg/l)	21.63	26.09	36.66	55.84	0.96	0.258
Sodium (mg/l)[c]	26.33	29.57	37.23	51.14	0.68	0.740
Magnesium (mg/l)[c]	17.57	20.23	26.51	37.90	0.98	0.067
Nitrate (mg/l)	1.07	0.85	0.32	0.00	0.88	0.038
Sulphate (mg/l)	42.51	50.71	70.12	105.33	0.03	1.000
Ammonia (mg/l)	0.29	0.21	0.00	0.00	0.07	0.382
Phosphate (mg/l)	0.62	0.52	0.30	0.00	0.68	0.075
COD (mg/l)	24.46	24.87	25.85	27.62	0.85	0.350
DO (mg/l)[d]	5.83	5.36	4.26	2.25	0.99	0.003
DOC (mg C/l)[c]	10.50	10.02	8.89	6.83	0.13	0.049

Notes: [c]In terms of the national water quality guidelines for domestic use.
[d]In terms of the International European Union water quality standards.

From Table 5, we can conclude that the water in the Grootdraai Dam catchment will become more alkaline should there be a gradual increase in the extent of urban built-up and mining land cover according to the predicted pH values. Furthermore, owing to an increase in mining operations, the catchment is also expected to experience an increase in its sulphate concentrations, which will reach tolerable standards in 2020. Certain water quality parameters such as nitrate, ammonia, phosphate, as well as DOC, are expected to decline in tandem with these developments. The complete transformation of the 'other' land cover class into urban built-up, mining, as well as cultivated land cover, would cause the catchment to record unacceptably high concentration levels of pH, EC, alkalinity, sulphate and DO, but also tolerable standards of chloride. In turn, these would lead to various environmental impacts and might cause economic growth and development to occur at a slower pace in future. The factors behind such a scenario would be the inhibition of certain industrial, mining and agricultural activities on account of scaling and the high alkalinity levels of the water, thereby precluding its use.

Conclusions

South Africa has limited natural water resources and is water stressed. The Vaal River catchment is part of the Vaal River system, which is deemed to be the 'workhorse' of South Africa as it is located within the economic heart of the country. The Grootdraai Dam catchment located within this Vaal River system helps supply water resources to all the major economic activities within the economic hub of the country and is fully exploited in terms of its water availability. The status of water quality within the catchment is therefore an important characteristic that needs to be investigated and monitored extensively due to its prime importance for the economic growth of the country. The degradation of the region's water quality will consequently decrease the availability of water in the catchment and have widespread environmental, social, as well as economic consequences and impacts.

Some relationships of varying significance were found between land cover change and the identified significant water quality parameters and it supports previous findings which illustrate that a change in land cover will be associated with hydrological responses but it affects water quality parameters at varying degrees and with different reactions. The hydrological changes associated with a specific water quality parameter will be dependent on the physical characteristics of the catchment as well as on the type of land cover and land use in terms of land use. These characteristics will in turn determine the type of hydrological response in terms of positive or negative relationships, the significance of the relationships and, lastly, which significant water quality parameters will change the most.

The establishment of these relationships has enabled the quantification of water quality change in the event of an increase or decrease of specific land cover classes. The future water quality conditions can therefore be predicted with the use of the formulated PLS model equations according to land cover change.

The future water quality of the Grootdraai Dam catchment and the Vaal River catchment as a whole will be affected significantly with the approval of the numerous prospected mines and the accompanied urban built-up areas within the catchment. This research therefore concludes that even the gradual increase of these land cover classes will have significant negative effects on the catchment's water quality and should be carefully considered or managed. In the event of the approval of these prospected mines, the catchment will experience tolerable to unacceptable standards of EC, alkalinity, nitrate, sulphate, phosphate COD and DO within the next couple of years. The tolerable and unacceptable concentrations of these water quality parameters will have widespread effects in terms of the environment and may inhibit future economic growth due to the water being of an unusable nature.

In the event of a rapid increase of mining land cover, which will be associated with the approval of the prospected mines, it will have even more significant and disastrous effects on the region's environment and economic spheres. It is therefore of prime importance that these developments be carefully considered if the region is to avert a water predicament.

The water quality of the Grootdraai Dam catchment as well as the Vaal River catchment as a whole can therefore be improved upon. The implementation of regular monitoring, creation of water-quality and land-cover databases, which in turn will promote and facilitate improved water management practices, informed decision-making processes as well as cost-effective measures can as a result promote the sustainability of the Vaal River catchment as a whole. To avert a predicament in the not too distant future, the approval of the prospected mines, especially in the Grootdraai Dam catchment, needs to be reconsidered and re-evaluated as these developments will have significant and detrimental environmental, social and economic effects on the region as well as on the country. The development of a science–policy interface will also assist greatly by promoting informed decision-making and can aid in the aversion of a water predicament.

Acknowledgements

The authors would like to thank the South African Weather Service for rainfall data, the Department of Water Affairs for evaporation and water flow data, Rand Water for water quality data, the Agricultural Resource Centre for 1994, 2000 and 2005 land cover, and the South African National Biodiversity Institute for the 2009 land cover dataset.

References

ARC (Agriculture Resource Centre). (2005). *Land cover for South Africa*. Pretoria, South Africa: ARC.

Ashton, P. J., & Haasbroek, B. (2002). *Water demand management and social adaptive capacity: A South African case study*. Johannesburg: CSIR: Division of Water, Environment and Forestry Technology.

Ashton, P. J., Hardwick, D., & Breen, C. M. (2008). Changes in water availability and demand within South Africa's shared river basins as determinants of regional social–ecological resilience. In M. J. Burns & A. V. B. Weaver (Eds.), *Exploring sustainability science: A Southern African perspective*, 279–310. Stellenbosch: Stellenbosch University Press.

Attua, E. M., Ayamga, J., & Pabi, O. (2014). Relating land use and land cover to surface water quality in the Densu River basin, Ghana. *International Journal of River Basin Management*, *12* (1), 57–68. doi:10.1080/15715124.2014.880711

Baker, B. (2011). *A decision support tool for predicting water quality based on land cover* (Master's Thesis). Nicholas School of the Environment of Duke University, Durham, NC.

Bertasso, A. (2004). *Ecological parameters of selected helminth species in Labeobarbus aeneus and Labeobarbus kimberleyensis in the Vaal Dam, and an evaluation of their influence on indicators of environmental health* (Unpublished MSc short dissertation). Rand Afrikaans University, Johannesburg, South Africa.

Burrough, P. A. (1990). *Principles of geographical information systems for land resource assessment*. Oxford, UK: Clarendon Press.

Chai, T., & Draxler, R. R. (2014). Root mean square error (RMSE) or mean absolute (MAE)? – Arguments against avoiding RMSE in the literature. *Geoscience Model Development*, *7*, 1247–1250. doi:10.5194/gmd-7-1247-2014

DEAT (Department of Environmental Affairs and Tourism). (2005). *National State of the Environment Report*. Pretoria: Government Printer.

Deelstra, J., Oygarden, L., Blankenberg, A. B., & Eggestad, H. O. (2011). Climate change and runoff from agricultural catchments in Norway. *International Journal of Climate Change Strategies and Management*, *3*, 345–360. doi:10.1108/17568691111175641

DWAF (Department of Water Affairs and Forestry). (2004). *Internal Strategic Perspective: Upper Vaal Water Management Area. Compiled by PDNA, WRP Consulting Engineers (Pty) Ltd, WMB and Kwezi-V3 on behalf of the Directorate: National Water Resource Planning, 2004* (Report No. P WMA 08/000/00/0304). Department of Water Affairs and Forestry, South Africa

Hall, M., Germain, R., Tyrrell, M., & Sampson, N. (2008). *Predicting future water quality from land use change projections in the Catskill–Delaware watersheds*. New York, NY: The State University of New York College of Environmental Science and Forestry and the Global Institute of Sustainable Forestry Yale University School of Forestry and Environmental Studies.

Helmschrot, J., & Flugel, W. A. (2002). Land use characterisation and change detection analysis for hydrological model parameterisation of large scale afforested areas using remote sensing. *Physics and Chemistry on the Earth*, *27*, 711–718. doi:10.1016/S1474-7065(02)00055-4

Huang, J., & Klemas, V. (2012). Using remote sensing of land cover change in coastal watersheds to predict downstream water quality. *Journal of Coastal Research*, *283*, 930–944. doi:10.2112/JCOASTRES-D-11-00176.1

Ibrahim, N., & Wibowo, A. (2013a). Predictions of water level in Dungun River Terengganu using partial least squares regression. *International of Basic and Applied Sciences*, *12*, 1–7.

Ibrahim, N., & Wibowo, A. (2013b). Partial least squares regression based variables selection for water level predictions. *American Journal of Applied Sciences*, *10*, 322–330. doi:10.3844/ajassp.2013.322.330

Jung, K., Lee, S., Hwang, H., & Jang, J. (2008). The effects of spatial variability of land use on stream water quality in a costal watershed. *Paddy Water Environment*, *6*, 275–284. doi:10.1007/s10333-008-0122-1

Kadewa, W., Moyo, B. H. Z., Mumba, P., & Phiri, O. (2005). Assessment of the impact of industrial effluent on water quality of receiving rivers in urban areas of Malawi. *International Journal of Environmental Science and Technology*, *2*, 237–244. doi:10.1007/BF03325882

King, N. A., Maree, G., & Muir, A. (2009). Freshwater systems. In H. A. Strydom & N. D. King (Eds.), *Fuggle & Rabie's Environmental management in South Africa* (2nd ed.). Johannesburg: Juta.

Li, S., Gu, S., Lui, W., Han, H., & Zhang, Q. (2008). Water quality in relation to land use and land cover in the upper Han River Basin, China. *Catena*, *75*, 216–222. doi:10.1016/j.catena.2008.06.005

Lo, C. P., & Yeung, A. K. W. (2002). *Concepts and techniques of geographic information systems: Chapter 4*. Upper Saddle River, NJ: Prentice Hall.

Longley, P. A., Goodchild, M. F., Maguire, D. J., & Rhind, D. W. (2005). *Geographic information systems and science: Chapter 6* (2nd ed.). Hoboken, NJ: Wiley.

Lou, B., Zhao, Y., Chen, K., & Zhao, X. (2009). Partial least squares regression model to predict water quality in urban water distribution. *Systems Transactions Tianjin University*, *15*, 140–144. doi:10.1007/s12209-009-0025-2

McCarthy, T. S. (2011). The impact of acid mine drainage in South Africa. *South African Journal of Science*, *107*, 1–7. doi:10.4102/sajs.v107i5/6.712

Meador, M. R., & Goldstein, R. M. (2003). Assessing water quality at large geographic scales: Relations among land use, water physicochemistry, riparian condition and fish community structure. *Environmental Management*, *31*, 504–517. doi:10.1007/s00267-002-2805-5

Nasser, F., & Wisenbaker, J. (2003). A Monte Carlo study investigating the impact of item parceling on measures of fit in confirmatory factor analysis. *Educational and Psychological Measurement*, *63*, 729–757. doi:10.1177/0013164403258228

Rothenberger, M. B., Burkholder, J. M., & Brownie, C. (2009). Long-term effects of changing land use practices on surface water quality in a coastal river and lagoonal estuary. *Environmental Management*, *44*, 505–523. doi:10.1007/s00267-009-9330-8

SANBI (South African Biodiversity Institute). (2009). *Land cover for South Africa*. Pretoria, South Africa: SANBI.

Seeboonruang, U. (2012). A statistical assessment of the impact of land uses on surface water quality indexes. *Journal of Environmental Management*, *101*, 134–142. doi:10.1016/j.jenvman.2011.10.019

Singh, A., Jakubowski, A. R., Chidister, I., & Townsend, P. A. (2013). A MODIS approach to predicting stream water quality in Wisconsin. *Remote Sensing of Environment*, *128*, 74–86. doi:10.1016/j.rse.2012.10.001

Thompson, M. W. (1996). A standard land cover classification scheme for remote sensing applications in South Africa. *South African Journal of Science*, *92*, 34–42.

Tong, S. T. Y., & Chen, W. (2002). Modeling the relationship between land use and surface water quality. *Journal of Environmental Management*, *66*, 377–393. doi:10.1006/jema.2002.0593

Usali, N., & Ismail, M. H. (2010). Use of remote sensing and GIS in monitoring water quality. *Journal of Sustainable Development*, *3*, 228–238. doi:10.5539/jsd.v3n3p228

Van Steenderen, R. A., Theron, S. J., & Hassett, A. J. (1987). The occurrence of organic micro-pollutants in the Vaal River between Grootdraai Dam and Parys. *Water S.A.*, *13*, 209–214.

Warburton, M. L., Schulze, R. E., & Jewitt, G. P. W. (2012). Hydrological impacts of land use change in three diverse South African catchments. *Journal of Hydrology*, *414–415*, 118–135. doi:10.1016/j.jhydrol.2011.10.028

Wilson, C., & Weng, Q. (2010). Assessing surface water quality and its relation with urban land cover changes in the Lake Calumet Area, Greater Chicago. *Environmental Management*, *45*, 1096–1111. doi:10.1007/s00267-010-9482-6

Assessing the existing knowledge base and opinions of decision makers on the regulation and monitoring of unconventional gas mining in South Africa

Surina Esterhuyse, Marthie Kemp and Nola Redelinghuys

A policy vacuum exists in relation to the exploration and mining of unconventional gas in South Africa, with a recent survey showing that 86% of the respondents did not know what hydraulic fracturing entails. We conducted a study to determine the opinion of decision makers involved in formulating policy and regulating mining activities related to shale gas mining in South Africa, as this was not covered in the aforementioned survey. Our results demonstrate that the regulation of shale gas mining in South Africa is viewed as extremely important and identifies possible regulatory and monitoring tools to assist in governing this activity.

Introduction

Unconventional gas mining by means of hydraulic fracturing is a new and unprecedented activity in South Africa that may result in a variety of impacts on both the socio-economic and biophysical environments. Shale gas mining and coalbed methane mining form part of unconventional gas mining, where gas reservoirs need to be stimulated in order to release the gas from the geological formations. This stimulation is usually achieved by means of hydraulic fracturing, although acidizing may also be used to stimulate gas production. Hydraulic fracturing entails the pumping of hydraulic fracturing fluid into a geological formation that contains gas and/or oil to increase its permeability. Various technologies can be combined or used separately during the hydraulic fracturing process. It may involve the use of only water (for water well stimulation) or a combination of any or all of four separate technologies, viz. directional drilling, the use of high volumes of fracturing fluids, the use of slickwater additives and the use of multi-well drilling pads. Hydraulic fracturing as used in the oil and gas industry commonly includes the usage of 0.5–2% chemical additives and large volumes of proppant (to keep the fracture zones that produce the oil or gas open), as well as large volumes of fluid (Broomfield, 2012).

South Africa is currently heavily dependent on fuel imports, which represented 21.4% of total merchandise imports in 2011 (World Bank, 2012). Production of fuel from unconventional gas could reduce the import bill significantly, making unconventional gas an attractive energy option. However, in pursuing unconventional gas, it is also important to consider the role that renewable energy resources may play in South Africa. In addition, water availability may significantly limit the expansion of energy sources (DOE, 2011), while the development of energy sources may also negatively impact on water security (Martin & Fischer, 2012).

Currently, a policy vacuum exists in relation to the exploration and mining of shale gas in South Africa (Havemann, 2011). However, these lacunae in the law do not exist only in South Africa. Even in the United States there is a drive to introduce fracking-specific legislation to areas that develop shale gas resources (Havemann, 2011). During 2012 at least 119 bills were introduced to address hydraulic fracturing in the United States (Pless, 2012), which illustrates the legal complexities related to regulating hydraulic fracturing.

A survey performed in South Africa by Ipsos Markinor for Shell in February 2012 found that 86% of respondents did not know what hydraulic fracturing entails (Harris & Fleetwood, 2012). This is problematic from the perspective of informed decision making, as decisions need to be based on knowledge of the issue at hand. However, the knowledge and opinions of decision makers involved in formulating policy and regulating mining activities related to shale gas mining were not covered in the mentioned survey.

Study objectives

This study aimed to identify the current knowledge base of decision makers regarding the impacts of shale gas mining by means of hydraulic fracturing and their opinions on the regulation of shale gas mining in South Africa. To meet this aim the following objectives were set:

- Determining the existing knowledge base of government regulatory entities and key stakeholders regarding shale gas mining and hydraulic fracturing
- Determining key informants' views on the importance of regulating and monitoring specific environmental and social impacts related to shale gas mining and hydraulic fracturing
- Determining key informants' views on South Africa's capacity to regulate and monitor the impacts related to shale gas mining
- Determining possible regulatory and monitoring tools that can assist in governing this activity and identifying possible regulatory bodies that should be responsible for the execution of various activities
- Identifying the departments or entities that, according to key informants, should primarily be responsible for specific regulatory activities
- Testing key informants' opinion on whether South Africa will be able to effectively regulate shale gas mining.

Methodology

Data was gathered by means of a structured questionnaire, administered via email.

A purposive sample was drawn that aimed to include a wide variety of stake-holders from relevant government departments, scientific and academic institutions, non-governmental organizations and academia. Thus, the intent was not to obtain a generalizable sample, but rather to purposefully select institutions and key informants within these institutions who could ultimately play an integral part in decision making on shale gas mining in South Africa. The institutions purposively targeted were the South African

Departments of Water Affairs (DWA), Mineral Resources (DMR), Environmental Affairs (DEA), and Agriculture, Forestry and Fisheries (DAFF), the Petroleum Association of South Africa (PASA), the Council for Geoscience (CGS), the Council for Scientific and Industrial Research (CSIR), the Centre for Environmental Rights (CER), the South African Earth Observation Network (SAEON), the World Wildlife Fund (WWF), the University of Cape Town (UCT), the University of the Free State (UFS), Nelson Mandela Metropolitan University (NMMU), and the Tswane University of Technology (TUT), as well as independent consultants in the field of environmental management. From these targeted institutions key informants were then purposively targeted, based on their knowledge and expertise on shale gas mining in South Africa. The majority of key informants (76%) had more than five years' experience in their respective fields.

Fifty-two questionnaires were sent out, and 25 respondents completed the questionnaire. A number of respondents declined to complete the questionnaire, citing sensitivity to the nature of the questions. Respondents were fairly evenly distributed among the contacted organizations. Twenty per cent of respondents came from the Department of Water Affairs, another 20% from consultants; 16% came from the Department of Environmental Affairs, and 12% each from science councils, academic institutions and the Department of Agriculture, Forestry and Fisheries. Disappointingly, only 4% of respondents came from the Petroleum Association of South Africa, which is a key institution in providing licenses for shale gas exploration and the regulation of petroleum related activities in South Africa.

After the questionnaires were received back, the completed questionnaires were coded and data were analyzed descriptively with the aid of IBM SPSS Statistics (version 20).

Results and discussion

The results will be discussed by looking at respondents' extent of knowledge on hydraulic fracturing, respondents' perceptions of important aspects to be monitored, the perceived capacity to deal with the impacts of hydraulic fracturing, the tools to help address the regulation of shale gas mining, and respondents' opinion about South Africa's ability to effectively regulate this activity.

Extent of knowledge on hydraulic fracturing

Although 60% of respondents indicated that they have extensive knowledge on the environmental impacts of mining in general, only 28% of respondents indicated extensive knowledge on the impacts of shale gas mining, and 36% on the environmental impacts of hydraulic fracturing specifically. When asked about their extent of knowledge on the regulation of shale gas mining internationally and in South Africa, 52% of respondents indicated limited knowledge on environmental regulation of hydraulic fracturing internationally, and 44% indicated limited knowledge on hydraulic fracturing in South Africa.

These statistics indicate the unprecedented nature of unconventional gas mining by means of hydraulic fracturing in South Africa. They can be explained by the fact that for many of the decision makers in South Africa there was until recently no need to familiarize themselves with issues pertaining to monitoring and regulating this activity.

Respondents' satisfaction with their current level of knowledge on shale gas mining was also tested. Half of the respondents indicated that they are not satisfied with their current level of knowledge, while 46% indicated that they are somewhat satisfied. This indicates a definite need for more in-depth information on shale gas mining in order to properly regulate and monitor this activity in South Africa.

Table 1.   Sources of knowledge.

Type of knowledge resource		Extent to which respondents used various knowledge sources		
		Not at all	Somewhat	Large extent
Scientific resources	Government reports	$n = 10$ 41.7%	$n = 5$ 20.8%	$n = 9$ 37.5%
	Research reports	$n = 8$ 33.3%	$n = 7$ 29.2%	$n = 9$ 37.5%
	Scholarly articles	$n = 10$ 41.6%	$n = 7$ 29.2%	$n = 7$ 29.2%
	Combined percentage of respondents who used scientific sources	*38.9%*	*26.4%*	*34.7%*
Popular media Sources	Internet sources	$n = 5$ 20%	$n = 8$ 32%	$n = 12$ 48%
	Printed media	$n = 3$ 12.5%	$n = 11$ 45.8%	$n = 10$ 41.7%
	Verbal media	$n = 11$ 45.8%	$n = 10$ 41.7%	$n = 3$ 12.5%
	Visual media	$n = 8$ 34.8%	$n = 8$ 34.8%	$n = 7$ 30.4%
	Talks and presentations	$n = 8$ 33.3%	$n = 9$ 37.5%	$n = 7$ 29.2%
	Other	$n = 3$ 30%	$n = 2$ 20%	$n = 5$ 50%
	Combined percentage of respondents who used popular media sources	*29.4%*	*35.3%*	*35.3%*

The various sources used by respondents to inform their knowledge on shale gas mining can be seen in Table 1. The usage of popular media sources was slightly higher than the usage of scientific sources, which can be a function of the lack of availability of scientific resources on unconventional gas mining and hydraulic fracturing, specifically pertaining to the South African situation. Various scientific reports reflect on the polarized nature of scientific data sources (DMR, 2012) and also question the validity of certain scientific reports (Broomfield, 2012). Popular media resources are not the most reliable data sources, since the media often rely on subjective views fuelled by public opinion. In the case of shale gas mining, the lack of scientific data at this stage creates a fertile breeding ground in the media for stirring sentiments and emotions on the issue. From the data it is clear that it is very difficult to come by reliable scientific resources, compelling decision makers to rely on less objective media sources.

Important aspects to monitor and regulate during shale gas mining, as perceived by respondents

Participants were asked to rate which aspects they regard as important to monitor and regulate during shale gas mining activities. The aspects that were rated by respondents as most important are reflected in Table 2.

All key informants indicated as important the monitoring of contamination arising from hydraulic fracturing, the monitoring of volumes of water used during unconventional gas mining, and the regulation of the disclosure of chemicals used during hydraulic fracturing.

Table 2. Important aspects to monitor and regulate during shale gas mining, as perceived by respondents.

Percentage of respondents who indicated an aspect as important to regulate and/or monitor	Activity to regulate and/or monitor
100	Regulating contamination arising from hydraulic fracturing Monitoring volumes of water usage Regulating the disclosure of chemicals
96	Regulating and monitoring land use planning regarding conservation areas Regulating the construction of gas mining wells to be compliant with proper construction standards Monitoring air quality impacts from shale gas mining Regularly testing well integrity during repeated fracking of the same wells Establishing baseline water quality before allowing hydraulic fracturing Monitoring the usage of fracking fluids
92	Determining the location of seismogenic (earthquake) zones to avoid drilling into seismically active zones that could trigger possible earthquakes Monitoring biodiversity loss
88	Regulating the use of "green" chemicals
84	Monitoring changes in the disease burdens of communities affected by shale gas mining Regulating and monitoring infrastructure development
80	Monitoring spatial development regarding urbanization Ensuring that jobs created are filled by South Africans

In a water-stressed country such as South Africa, it is not surprising that the regulation and monitoring of water-related issues are seen as paramount. In the drier parts of the US, and in dry countries such as Australia, the water volume used for hydraulic fracturing is also a key issue (Australian National University [ANU], 2012; Rahm, 2011; Williams, Stubbs, & Milligan, 2012). Sourcing of water for fracturing operations may have large impacts on the quality and quantity of both surface water and groundwater (Lechtenböhmer et al., 2011; Rahm & Riha, 2012; USEPA, 2011b; Williams et al., 2012).

Authors such as Rahm (2011), Furlow and Hays (2012), Sakmar (2011) and Pless (2012) all identify the disclosure of chemicals used in the hydraulic fracturing process as important. In practice, however, hydraulic fracturing chemicals are not disclosed consistently. For example, the state of Colorado (USA) requires disclosure of chemicals and concentrations added to fracking fluids to physicians and regulators only in an emergency event, thereby preserving drillers' trade secrets, while Wyoming requires full public disclosure of chemical additives (Rahm, 2011). In 2011, House Bill 3328 (Texas Legislature Online, 2011) was passed in Texas – the first hydraulic fracturing disclosure legislation and regulation specific to hydraulic fracturing in Texas (Furlow & Hays, 2012). The US Congress introduced in 2009 the Fracturing Responsibility and Awareness Act (H.R. 1084) (FRAC Act) (United States Congress, 2009), which aims to define hydraulic fracturing as a federally regulated activity under the United Sates Safe Drinking Water Act (Furlow & Hays, 2012; Sakmar, 2011). South Africa would be prudent to draft regulations that require disclosure of chemicals used in fluids during drilling or fracturing operations to relevant institutions.

Establishing baseline water quality before allowing hydraulic fracturing was identified as important by 96% of the participants, indicating that key informants are aware of the window of opportunity to establish a monitoring baseline in terms of hydraulic fracturing. This activity is very important if landowners want to determine whether groundwater contamination is due to unconventional gas mining operations or to another cause. In the US, the states of Colorado, Ohio and Pennsylvania require that operators conduct baseline water testing (Government Accountability Office [GAO], 2012a).

Cases of alleged groundwater contamination in the US could usually not be proven because officials cannot link changes in groundwater quality to oil and gas activities, and can thus take no legal action. This is often due to the fact that no baseline data exist on the quality of groundwater prior to oil and gas development (GAO, 2012a, 2012b). In South Africa, the Department of Water Affairs, as the custodian of South Africa's water, should take the lead to proactively protect South Africa's water resources on a national level, and this includes performing a baseline survey of water quality in areas where unconventional gas may be mined in future.

Regarding the protection of water sources, key informants deemed as important the regulating of well integrity on a regular basis and the regulating of the use of fracking fluids, including "green" chemicals. Internationally, mechanical failure and deformation of wells represent widespread diffuse sources of water pollution over the long term in gas mining areas (Bishop, 2011; Dusseault, Gray & Nawrocki, 2000), while well abandonment and the poor sealing of wells after well decommissioning may lead to long-term groundwater contamination legacy issues (ANU, 2012; Broomfield, 2012; National Research Council [NRC], 2012b). Wastewater treatment may also pose challenges in terms of brine management (ANU, 2012), and if wastewater is re-injected into deeper porous geological formations it may cause geological and aquifer deformation, with the associated possibility of triggered seismicity (Lechtenböhmer et al., 2011; NRC, 2012a; Zoback, Kitasei, & Copithorne, 2010) and possible fluid migration (Broomfield, 2012; USEPA, 2011b). Deteriorating water quality would also impact negatively on the health of communities (Broderick, Anderson, Wood, Gilbert, & Sharmina, 2011; Coburn et al., 2011).

Overall, it can be concluded that respondents are very much aware of the importance of proper monitoring and regulation of the potential environmental impacts of shale gas mining. Monitoring different aspects of unconventional gas mining before, during and after mining is important for almost all countries where unconventional gas mining may occur (ANU, 2012; Ohio Environmental Law Centre [OELC], 2012).

Capacity to manage the impacts of unconventional gas mining by means of hydraulic fracturing

Respondents were asked to share their opinion regarding the capacity of South Africa to deal with various aspects of unconventional gas mining. Statements were given which the respondents had to rate on a scale from 1 (not accurate at all) to 5 (completely accurate). The most pertinent responses can be viewed in Table 3.

Ninety-four per cent of respondents viewed the Mineral and Petroleum Resources Development Act (Act No. 28 of 2002) (MPRDA) (Republic of South Africa [RSA], 2002) as insufficient in scope to manage the wide range of challenges presented by unconventional gas mining and fracking. This view has been confirmed by legal minds in South

Table 3. Capacity to regulate unconventional gas mining by means of hydraulic fracturing in South Africa.

	Percentage of respondents	Statements given in the questionnaire regarding capacity to regulate shale gas mining in South Africa
Respondents agreeing with statement	92	Regulation of possible impacts is complex due to conflicting mandates of different departments.
	82	Uncertainties exist between local and national government departments on responsibilities for monitoring and regulation of different fracking aspects.
		Regulation complexities exist due to fragmentation of responsibilities within specific departments.
	76	Mining rights authorization processes in South Africa are fragmented.
	64	Amendments are required in terms of South African statutes to make development of fracking-specific regulations possible.
	48	Mining rights authorization processes in South Africa are limited.
Respondents disagreeing with statement	94	South Africa possesses sufficient fracking-specific legislation and fracking-specific policies.
		The Mineral Petroleum Resources Development Act (1998) is sufficient in scope to deal with the challenges of fracking.
		South Africa has sufficient institutional capacity to monitor shale gas mining operations.
	88	South Africa has sufficient institutional capacity to enforce compliance with conditions of license approval for gas mining operations.
	76	Information on the potential environmental and health risks of fracking is currently sufficient.

Africa (Havemann, 2011; Havemann, Glazewski, & Brownlie, 2011; Kantor, 2011a), who state that extensive regulations need to be drafted under the MPRDA and other acts to address fracking sufficiently.

In comparison, the South African water law and policy framework is one of the most progressive worldwide. The development of this policy framework has been initiated by the water reform programme in South Africa, which included the revision of policy and the drafting of new legislation to address issues of environmental sustainability and the efficient use of water (Schreiner, 2012). However, an issue that hampers the implementation of the legislation in practice is overly complex systems for the implementation of legislation, making the total demand on skilled resources in the water sector too large to sufficiently handle (Schreiner, 2012). This may explain why respondents are of the opinion that South Africa has insufficient institutional capacity to monitor and enforce compliance with conditions of license approval for shale gas mining operations.

A further concern highlighted is the perceived conflict in mandates between different South African government departments that may hamper effective regulation, as indicated by 92% of respondents. For example, environmental controls for mining legislation are at present still the responsibility of the Minister of Mineral Resources (not the Minister of Environmental Affairs), which presents a conflict of interest (Kantor, 2011b). This situation may be corrected in due course.

Tools to help address the regulation of shale gas mining

Tools or actions which may assist authorities with regulating unconventional gas mining activities were presented to respondents, who had to indicate which may be of the most help in regulating unconventional gas mining activities in South Africa. The responses are reflected as the percentage of respondents who indicated a tool or activity as "extremely useful" and can be seen in Table 4.

All the key informants viewed as useful the development of fracking-specific monitoring protocols before allowing hydraulic fracturing to commence. Developing monitoring protocols before allowing hydraulic fracturing is important for the monitoring of baseline water quality specifically, as well as other baselines (biophysical aspects and socio-economic aspects). Monitoring baseline conditions before allowing unconventional gas mining by means of hydraulic fracturing is a very important aspect if citizens want to protect their property and the environment (ANU, 2012; GAO, 2012a, 2012b). Additionally, monitoring efforts should be coordinated regionally in South Africa to ensure consistency in monitoring activities or protocols. These activities should ideally be managed on a national level, and the coordination role should be the responsibility of the relevant national government authorities (e.g. the DWA for water). This would require strategic planning on a national scale.

Table 4. Tools to help address the regulation of shale gas mining.

Percentage of respondents regarding this tool as extremely useful	Tools to help address the regulation of shale gas mining
100	Developing fracking-specific monitoring protocols before allowing hydraulic fracturing (HF) to commence
96	Performing research to identify and assess the impacts related to HF before giving the go-ahead for exploration
92	Developing fracking-specific legislation before allowing HF to commence Declaring HF a controlled activity under s. 38 of the National Water Act Establishing a central database to store fracking-related information for ease of access and centralized management Identifying and clearly specifying the mandates, roles and responsibilities of local government versus national government
84	Performing a Strategic Environmental Assessment instead of an Environmental Impact Assessment to determine potential impacts Performing a detailed strategic assessment of the available energy generation options in South Africa before deciding on allowing HF
72	Requiring oil and gas companies to give security under s. 30 of the National Water Act for the protection of the water resource or property in respect of any obligation or potential obligation arising from a license to be issued Establishing an independent entity to monitor fracking activities and report to government on a regular basis
60	Adopting management policies from other countries where HF is currently taking place
52	Enforcing the self-regulation and reporting to government of oil and gas companies as part of their license conditions Placing an indefinite moratorium on fracking if other sources of energy are found to be sufficient

Ninety-six per cent of participants felt that performing research to identify and assess the impacts related to hydraulic fracturing before allowing it to commence would be useful. This is highlighted as important by the legal fraternity, before an effective legislative framework can be drafted for hydraulic fracturing (Havemann, 2011; Kantor, 2011a, 2011b). Research on hydraulic fracturing seems to be limited at the moment. The study done by the parliamentary task team on unconventional gas mining in South Africa (DMR, 2012) was possibly the most comprehensive research done on the issue to date. However, the process followed during the drafting of the DMR report by the task team has been criticized by Kantor (2011a) for not being transparent, which may render these research results questionable. The DMR research report was also drafted over a period of 12 months, while a country such as the US, where hydraulic fracturing has been performed for over a decade, plans to identify possible hydraulic fracturing impacts over a period of 3 years (USEPA, 2011a).

A worrying development regarding the pursuit of unconventional gas mining in South Africa was that decision makers deemed this report to be sufficient evidence to lift a moratorium on the acceptance and processing of unconventional oil and gas mining applications by PASA. Internationally, the trend is for countries to keep their moratoria in place until such time as proper and transparent scientific investigations on possible impacts have been performed and fracking-specific legislation and regulations, based on scientific results, have been drafted (Philippe & Partners, 2011; Pless, 2012).

Most of the participants (92%) agreed that developing fracking-specific legislation and regulations before allowing hydraulic fracturing to commence would be important. This finding is supported by other authors (Kantor, 2011a; Havemann, 2011; Havemann et al., 2011), while the parliamentary task team report (DMR, 2012) also prudently advised government that exploration may be allowed, but hydraulic fracturing should not, until such time as legislation and regulations have been drafted to effectively address hydraulic fracturing. Other tools that were also specified as important by 92% of the participants included declaring hydraulic fracturing a controlled activity under s. 38 of the National Water Act (Act No. 36 of 1998) (NWA) (RSA, 1998), establishing a central database to store fracking-related information, and identifying and clearly specifying the mandates, roles and responsibilities of local government versus national government.

A way to assess potential impacts of unconventional gas mining in a coherent manner across South Africa would be to perform strategic environmental assessments (SEAs) instead of environmental impact assessments (EIAs). Conducting SEAs was indicated as important by 84% of respondents. Environmental impact assessments may be insufficient in the case of unconventional gas mining, since impacts related to unconventional gas mining occur cumulatively on a regional scale and may pose legacy issues (ANU, 2012).

Interestingly, performing a strategic assessment of available energy generation options is viewed as important by only 84% of participants. This is an important first activity to be performed by government to guide decision making on allowing unconventional gas (a non-sustainable fossil fuel) as an energy option. Alternative power generation options should be compared to unconventional gas resources in a cost-benefit analysis, while also factoring in environmental and socio-economic costs. Politicians and decision makers should understand the water–energy–food nexus as it applies to South Africa so that proper decisions are made to ensure the long-term sustainable use of South Africa's resources. The National Development Plan (National Planning Commission [NPC], 2012) recommends fast-tracking of gas-to-power projects if shale and coalbed methane gas reserves are proven and environmental concerns alleviated (NPC, 2012). This plan recommends exploratory drilling to identify economically recoverable gas reserves while environmental investigations continue to ascertain whether sustainable exploitation of these resources is possible. Performing a strategic assessment of energy generation options available to South Africa would tie in well with the broad framework of this plan.

Only 60% of participants felt that adopting policies from other countries where hydraulic fracturing is currently allowed may be useful. Adoption of policies and regulations from other countries needs to be performed with caution, since the potentially negative consequences of fracking may vary depending on the particularities of locations in which fracking is proposed (Havemann, 2011). Nationally, South Africa presents unique paleontological and astronomy-related concerns that oil and gas applicants would not encounter in, for instance, the United Kingdom. Large geographic areas with features suitable for optical and radio astronomy gives South Africa a unique astronomical advantage, as is highlighted by the Astronomy Geographic Advantage Act (Act No. 21 of 2007) (RSA, 2007), which serves to protect astronomy-related activities in South Africa. From a local perspective, each well pad and the associated infrastructure will have impacts that affect the unique environment in a particular area (e.g. localized impacts on vegetation, faunal populations, archaeology and socio-economics). Detailed consideration should be given to the numerous amendments that would need to be made to the many statutes and regulations to be affected by the introduction of fracking-specific regulations, which makes the simple adoption of a foreign fracking-specific regulatory regime inappropriate.

Tools that received less support from respondents included "enforcing the self-regulation and reporting to government of oil and gas companies as part of their license conditions" and "placing an indefinite moratorium on fracking if other sources of energy are found to be sufficient" (both 52%). Self-regulation may be a useful tool if government does not have sufficient manpower or capacity to perform monitoring, but the trustworthiness of oil and gas companies to self-regulate remains questionable (International Energy Agency [IEA], 2012; Nwokocha, Ugwu, Fagbenro, & Cookey, 2012; Ten Kate, 2011). Participants also felt that placing an indefinite moratorium on oil and gas mining, if other sources of energy are found to be sufficient, may be a useful course of action. However, this tool did not receive high support from the participants, possibly due to the fact that South Africa has been so dependent on fossil fuels in the past, and because anti-nuclear lobbying may cause government anxiety.

In conclusion, the most important tools that may help in the regulation of shale gas mining have been identified by participants, and these tools are in line with international trends. Stemming from the concern over the anticipated impacts of hydraulic fracturing on water specifically, it is crucial that specific attention be given to the development of water policy, both internationally and in South Africa. Proper water-related policy is especially important since the country is water scarce and water serves as both a driver and limiter of economic development (Blignaut & van Heerden, 2009; DWA, 2012b). Some of the water-related policy issues were highlighted in the questionnaire, and it was clear from the responses that proper water management in South Africa is very important.

Identification of the departments primarily responsible for specific regulatory activities

Respondents were asked to identify which departments should take primary responsibility for specific activities. Respondents prefer departments to share responsibility for developing fracking-specific legislation according to their respective mandates. Respondents deemed the Department of Water Affairs to be responsible for developing water-related policy and monitoring water consumption and pollution; the Department of Environmental Affairs to be responsible for the monitoring of biodiversity impacts, air quality impacts, vegetation loss and the impacts of fracking on land use; and the Department of Mineral Resources to be the main authority responsible for approving mining applications.

The approval of mining applications is currently governed by mining legislation under the auspices of the DMR. Other countries operate in a similar fashion, in that mining legislation fulfils a central role in governing the authorization and permitting procedures for exploration or production of hydrocarbons (Philippe & Partners, 2011).

Interestingly, 20–36% of participants were of the opinion that an independent entity or organization should be established to oversee the regulation and monitoring of hydraulic fracturing. Perhaps this is something to consider, to ensure that these tasks are performed in a coordinated and transparent manner.

Opinion on whether South Africa will be able to effectively regulate shale gas mining

Respondents were lastly asked to comment on South Africa's ability to effectively regulate shale gas mining. Seventy-two per cent were of the opinion that South Africa will not be able to effectively manage the challenges pertaining to shale gas mining, while 16% were unsure; only 12% thought that South Africa would be able to effectively manage these challenges. This opinion may be based on the fact that fracking-specific regulations do not currently exist in South Africa (Havemann, 2011; Havemann et al., 2011; Kantor, 2011a), and also on the fact that currently South Africa does not effectively manage existing mining applications. In a parliamentary reply by the Department of Water Affairs, it emerged that during 2011, 53 mines were still operating without water use licenses (DWA, 2012a). This fact points to poor coordination between the national Department of Mineral Resources and Department of Water Affairs, which should ensure that both mining permits and water use licenses are in place before mining operations can proceed.

Conclusion

From the results of the study it can be concluded that key informants have limited knowledge on shale gas mining and hydraulic fracturing, as well as on the regulation of this type of mining, in terms of their understanding of regulation both in different parts of the world and in South Africa. The fact that half of the key informants indicated that they are not at all satisfied with their current knowledge on shale gas mining is indicative of the unprecedented nature of this activity in South Africa and points towards a serious need for knowledge dissemination through which decision makers can be exposed to relevant, current and applicable information on shale gas mining and hydraulic fracturing, specifically with reference to South Africa. This conclusion is strengthened by the fact that more key informants relied on the use of popular media than on scientific sources to expand their knowledge base and highlights the urgency of providing more scientific, scholarly information to those in decision-making positions. Therefore, various avenues of dissemination of information will need to be explored in the South African context to assist decision makers in their decision-making processes.

Key informants are in general aware of the importance of proper monitoring and regulation of shale gas mining. It is telling that the monitoring of water-related issues was rated as the most important activity to be performed, especially in a water-scarce country such as South Africa, and this finding is supported by current literature on the water-related impacts of shale gas mining (Lechtenböhmer et al., 2011; Rahm & Riha, 2012; USEPA, 2011b). Thus, specific attention needs to be given to the development of applicable water policy to address these issues. Highlighted as most important were the monitoring of contamination arising from hydraulic fracturing, the monitoring of the volumes of water used, and the regulation of the disclosure of chemicals used in the fracturing process. Establishing a baseline on water quality before allowing hydraulic fracturing to take place was also strongly emphasized by respondents.

It is worrying that key informants are largely unconvinced that South Africa has sufficient capacity to regulate and monitor shale gas mining or to manage the challenges emanating from this type of mining. The overall majority of respondents indicated that South Africa has insufficient fracking-specific legislation and policy and that the MPRDA is currently insufficient in scope to manage the challenges represented by shale gas mining and hydraulic fracturing. This opinion may be based on the fact that fracking-specific regulations do not currently exist in South Africa (Havemann, 2011; Havemann et al., 2011; Kantor, 2011a) and also on the fact that currently South Africa does not succeed in effectively managing existing mining applications. Additionally, the majority of respondents identified a conflict in departments' mandates that may hamper effective regulation of this activity in South Africa. These findings reveal that regulators, academia and consultants are of the opinion that regulatory gaps exist that need to be addressed.

Among the tools that were identified to regulate shale gas mining, fracking-specific monitoring protocols were deemed the most important. Key informants also indicated that fracking-specific legislation and regulations should be developed before hydraulic fracturing is allowed to take place. With regard to the departments primarily responsible for specific regulatory activities, most respondents indicated that specific departments should assume their traditional responsibilities according to their mandates. However, a number of key informants opined that an independent entity should perform all of these tasks. This interesting finding correlates with the key informants' views on the existing capacity in the country to monitor and regulate shale gas mining. It could indicate that at least some of the respondents view the establishment of an independent monitoring agency as a way to mitigate the current lack of monitoring and regulation capacity in existing institutions. Based on this, the question of whether an independent organization should be established in South Africa to perform these tasks in a coordinated manner needs to be considered by those in decision-making positions.

South Africa is in the fortunate position that proactive steps can be taken now, before exploration and mining are fully pursued. Relevant government departments and institutions should apply the precautionary principle by allowing time for scientific investigation, identifying possible risks and drafting relevant legislation to prevent or minimize adverse effects of unconventional gas mining on the environment, before any of these impacts occur.

Acknowledgements

The authors would like to thank all the institutions and respondents that participated in the study, as well as students at the Centre for Environmental Management (Joan Adendorff, Esté Prinsloo and Arjen Nell) who assisted with data capture. We also appreciate the valuable suggestions of the anonymous reviewers, whose comments improved an earlier draft of this manuscript.

References

Australian National University. (2012). *Unconventional gas production and water resources – lessons from the United States on better governance – a workshop for Australian government officials*. Canberra: Crawford School of Public Policy.

Bishop, R. E. (2011). *Chemical and biological risk assessment for natural gas extraction in New York (Draft Report)*. New York, NY: State University of New York, College at Oneonta.

Blignaut, J., & van Heerden, J. (2009). The impact of water scarcity on economic development initiatives. *Water SA*, 35(4), 415–420. Retrieved from Scielo South Africa: http://www.scielo.org.za/scielo.php?pid=S1816-79502009000400006&script=sci_arttext

Broderick, J., Anderson, K., Wood, R., Gilbert, P., & Sharmina, M. (2011). *Shale gas: An updated assessment of environmental and climate change impacts*. Retrieved from Tyndall Centre: http://www.tyndall.ac.uk/sites/default/files/coop_shale_gas_report_update_v3.10.pdf

Broomfield, M. (2012). *Support to the identification of potential risks for the environment and human health arising from hydrocarbons operations involving hydraulic fracturing in Europe* (Report No. ED57281-17). Retrieved from European Commission: http://ec.europa.eu/environment/integration/energy/pdf/fracking%20study.pdf

Coburn, M., Kozikowski, T., Lloyd, N., Meredith, B., Parnes, D., Reed, J., & Shah, R. (2011). *Positives, negatives, uncertainty, and opinions on hydrofracturing in the Marcellus Shale*. Retrieved from Allegany County Government Services and Information: http://www.gov.allconet.org/pr/FRACKING%20RESEARCH%20PAPER%20FINAL-5-27-11.pdf

Department of Energy. (2011). *Integrated resource plan for electricity 2010–2030, Revision 2: Final Report*. South Africa: Author.

Department of Mineral Resources. (2012). *Report on investigation of hydraulic fracturing in the Karoo Basin of South Africa*. South Africa: Author.

Department of Water Affairs. (2012a). *National assembly: Question 468 for written reply*. Retrieved from Department of Water Affairs: http://www.dwaf.gov.za/communications/Q&A/2012/NA%20468.pdf

Department of Water Affairs. (2012b). *Proposed National Water Resource Strategy 2 (NWRS2) Summary – Managing water for an equitable and sustainable future*. South Africa: Author.

Dusseault, M. B., Gray, M. N., & Nawrocki, P. (2000, November). *Why oilwells leak: Cement behaviour and long-term consequences*. Paper presented at the Society of Petroleum Engineers International Oil and Gas Conference and Exhibition, Beijing, China.

Furlow, J. D., & Hays, J. R. (2012). Disclosure with protection of trade secrets comes to the hydraulic fracturing revolution. *Texas Journal of Oil, Gas and Energy Law, 7*, 289–355. Retrieved from http://web.ebscohost.com/ehost/pdfviewer/pdfviewer?sid=df554bb4-f38b-44e7-8e9a-8c9c3fe00f3c%40sessionmgr15&vid=1&hid=26

Government Accountability Office. (2012a). *Unconventional oil and gas development – key environmental and public health requirements*. Washington, DC: United States Government Accountability Office.

Government Accountability Office. (2012b). *Information on shale resources, development, and environmental and health risks*. Washington, DC: United States Government Accountability Office.

Harris, M., & Fleetwood, S. (2012). *Future energy report*. Pretoria: Ipsos Markinor.

Havemann, L. (2011). A South African regulatory framework for fracking: is the cart being put before the horse? *Mining Prospectus*. Retrieved from http://www.havemanninc.com/wp-content/uploads/2011/12/Mining-Prospectus-1.pdf

Havemann, L., Glazewski, J., & Brownlie, S. (2011). *A critical review of the application for a Karoo gas exploration right by Shell Exploration Company B.V.* Cape Town: Havemann Inc Specialist Energy Attorneys.

International Energy Agency. (2012). *Golden rules for a golden age of gas*. Paris: Author.

Kantor, P. (2011a). A cautious and risk-averse approach. *De Rebus, 105,* 32–34. Retrieved from http://www.myvirtualpaper.com/doc/derebus/de_rebus_december_2011/2011112401/34.html#34

Kantor, P. (2011b). *A legal framework for the debate*. Retrieved from http://www.sibergramme.co.za/blog/fracking-a-legal-framework-for-the-debate

Lechtenböhmer, S., Altman, M., Capito, S., Matra, Z., Weindrorf, W., & Zittel, W. (2011). *Impacts of shale gas and shale oil extraction on the environment and on human health*. (Report No. IP/A/ENVI/ST/2011–07). Policy Department (Economic and Scientific Policy). Retrieved from http://www.europarl.europa.eu/document/activities/cont/201107/20110715ATT24183/20110715ATT24183EN.pdf

Martin, B., & Fischer, R. (2012). *The energy-water nexus: Energy demands on water resources*. Retrieved from http://www.groundwork.org.za/ClimateHealthRoundtables/water-energy-nexus.pdf

National Planning Commission. (2012). *National Development Plan 2030 Our Future – make it work*. Retrieved from the South African National Planning Commission: http://www.

npconline.co.za/MediaLib/Downloads/Downloads/NDP%202030%20-%20Our%20future%20-%20make%20it%20work.pdf

National Research Council. (2012a). *Induced seismicity potential in energy technologies.* Washington, DC: The National Academies Press.

National Research Council. (2012b). *Prepublication copy: Alternatives for managing the nation's complex contaminated groundwater sites.* Washington, DC: The National Academies Press.

Nwokocha, C. O., Ugwu, D. U., Fagbenro, A. B., & Cookey, E. J. (2012). Environmental impacts of oil and gas production in Nigeria. *Journal of Energy and Power Engineering, 6,* 70–75. Retrieved from http://www.davidpublishing.com/davidpublishing/Upfile/2/16/2012/2012021605670340.pdf

Ohio Environmental Law Centre. (2012). *Fracking litigation report. Vol 1. Filling in the structural cracks of fracking regulation.* Retrieved from Ohio Environmental Law Centre: http://ohioenvirolawcenter.files.wordpress.com/2012/11/vol-1-filling-in-the-structural-cracks-of-fracking-regulation.pdf

Philippe & Partners. (2011). *Final report on unconventional gas in Europe.* Brussels: European Union.

Pless, J. (2012, June). *Natural gas development and hydraulic fracturing – a policymaker's guide.* Denver, CO: National Conference of State Legislatures.

Rahm, D. (2011). Regulating hydraulic fracturing in shale gas plays: The case of Texas. *Energy Policy, 39,* 2974–2981. doi: 10.1016/j.enpol.2011.03.009.

Rahm, B. G., & Riha, S. J. (2012). Towards strategic management of shale gas development: Regional, collective impacts on water resources. *Environmental Science and Policy, 17,* 12–23. doi: 10.1016/j.envsci.2011.12.004.

Republic of South Africa. (1998). *National Water Act No. 36 of 1998.* South Africa: Author.

Republic of South Africa. (2002). *Mineral and Petroleum Resources Development Act No. 28 of 2002.* South Africa: Author.

Republic of South Africa. (2007). *Astronomy Geographic Advantage Act No. 21 of 2007.* South Africa: Author.

Sakmar, S. L. (2011). The global shale gas initiative: Will the United States be the role model for the development of shale gas around the world?. *Houston Journal of International Law, 33(2),* 369–416. Retrieved from http://web.ebscohost.com/ehost/pdfviewer/pdfviewer?sid=5a81a0ce-96a7-4619-9f23-0eab41a6e10d%40sessionmgr4&vid=1&hid=26

Schreiner, B. (2012, November). *Opening and Welcome address by Barbara Schreiner at the International Conference on Freshwater Governance.* International Freshwater Governance Conference, Champagne Sports Resort, Drakensberg, South Africa.

Ten Kate, A. (2011). *Royal Dutch Shell and its sustainability troubles. Background report to the Erratum of Shell's Annual Report 2010.* Amsterdam: Milieudefensie.

Texas Legislature Online. (2011). *House Bill 3328.* Retrieved from: http://www.capitol.state.tx.us/BillLookup/Text.aspx?LegSess=82R&Bill=HB3328

United States Congress. (2009). *Fracturing Responsibility and Awareness Act (H.R. 1084).* Retrieved from The Library of Congress: http://thomas.loc.gov/cgi-bin/bdquery/z?d112:h.r.1084

USEPA. (2011a). *EPA submits draft hydraulic fracturing study plan to independent scientists for review/the draft plan is open to public comment.* Retrieved from United States Environmental Protection Agency: http://yosemite.epa.gov/opa/admpress.nsf/d0cf6618525a9efb85257359003fb69d/26195e235a35cb3885257831005fd9cd!OpenDocument

USEPA. (2011b). *Draft plan to study the potential impacts of hydraulic fracturing on drinking water resources.* (Report No. EPA/600/D-11/001/February 2011/www.epa.gov/research). Retrieved from United States Environmental Protection Agency: http://water.epa.gov/type/groundwater/uic/class2/hydraulicfracturing/upload/hf_study_plan_110211_final_508.pdf

Williams, J., Stubbs, T., & Milligan, A. (2012). *An analysis of coal seam gas production and natural resource management in Australia.* Canberra: John Williams Scientific Services Pty Ltd.

World Bank. (2012). *South Africa fuel imports (% of merchandise imports)* Retrieved from World Bank: http://search.worldbank.org/quickview?name=%3Cem%3EFuel%3C%2Fem%3E+imports+%28%25+of+merchandise+imports%29&id=TM.VAL.FUEL.ZS.UN&type=Indicators&cube_no=2&qterm=south+africa+fuel+import

Zoback, M., Kitasei, S., & Copithorne, B. (2010). *Addressing the environmental risks from shale gas development (briefing paper 1).* Washington, DC: Worldwatch Institute.

A water-energy-food security analysis tool for mining in Suriname: operationalizing the Mining Policy Framework of the Intergovernmental Forum on Mining, Minerals, Metals and Sustainable Development

Dimple Roy, Darren Swanson, Carter Borden, Alec Crawford, Livia Bizikova and Gabriel Huppe

Background

The Plan of Implementation coming out of the 2002 World Summit on Sustainable Development highlighted how the contribution of mining and minerals extraction to sustainable development could be enhanced (UN, 2002, para. 46). Motivated by the summit and its plan, a number of countries came together through a voluntary partnership to establish the Intergovernmental Forum on Mining, Minerals, Metals and Sustainable Development (IGF), and its secretariat was then established in 2005 within the Canadian government's international development agency.

The IGF provides a platform for national governments with an interest in mining to work together to advance the priorities identified at the 2002 summit.[1] It is the only global policy forum for the mining and metals sector with the overarching objective of enhancing capacity for good governance in the sector. The forum continues to expand: 55 countries are now members of the IGF, with France, Germany, the Islamic Republic of Iran, the Netherlands and Rwanda all joining in 2015. In 2015, the secretariat of the IGF was moved to the International Institute for Sustainable Development, a research policy think tank based in Canada. The institute had previously served as the regional partner for the global Mining, Minerals and Sustainable Development initiative in 2000, and in 2009 the institute was invited to join the core administrative group of the International Bar Association's Mining Law Committee, for which it prepared a Model Mining Development Agreement.

In 2013, the IGF finalized its Mining Policy Framework (MPF) to advance good governance in mining practices in its member states. The MPF provides a compendium of activities that the IGF member countries have identified as best practice "for exercising good governance of the mining sector and promoting the generation and equitable sharing of benefits in a manner that will contribute to sustainable development" (IGF, 2013). These best practice standards are organized along six key pillars:

- legal and policy framework
- financial benefit optimization
- socio-economic benefit optimization
- environmental management
- post-mining transition
- artisanal and small-scale mining.

Mines and mining activity have environmental and socio-economic benefits and impacts on the local communities and ecosystems in which they are located, with implications for national and regional well-being. While all MPF pillars work towards better governance, the pillars specifically addressing local impacts include environmental management, socio-economic benefit optimization, legal and policy framework, and artisanal and small-scale mining (ASM) (Table 1). Environmental management mostly addresses monitoring and reporting, and limiting the impacts of mining activities on water and biodiversity systems. Socio-economic benefit optimization addresses evaluating mining impacts on: community health; community integration within the local, regional and national fabric; and recognizing and respecting human rights, indigenous populations and cultural heritage. The legal and policy framework pillar involves establishing systems to involve local stakeholders, conducting integrated social, economic and environmental impact assessments, and respecting cultural heritage. And

Table 1. Pillars of the Mining Policy Framework relating to local impacts (IGF, 2013).

MPF pillars relating to local impacts	Activities
Environmental management	• Water: develop water management standards and requiring mining entities to ensure that the quality and quantity of mine effluent streams discharged to the environment, including storm water, leach pad drainage, process effluents, and mine works drainage, are managed and treated to meet established effluent discharge guideline values both on and off-site; • Biodiversity: require that mining entities submit environmental management programmes; identify, monitor and address potential and actual risks and impacts to biodiversity; and conduct continuous monitoring and compile and submit performance assessments.
Socio-economic benefit optimization	• Evaluate integration mines and mining into the local, regional and national fabrics; • Address community health • Create business development opportunities: • Address potential security issues; and • Respect human rights, indigenous peoples, and cultural heritage
Legal and policy framework	• Prepare applications for a mining permit, consult with communities and other stakeholders at stages of planning and development; • Submit integrated social, economic and environmental assessments; • Address indigenous peoples, cultural heritage, resettlement, and community safety and security issues in permit applications; • Provide affected communities with an opportunity to express their views on project risks and impacts and be consulted on the development of mitigation measures.
Artisanal and small-scale mining	• Improve savings in the artisanal mining community; • Improve the collection, management and reinvestment of ASM revenue; • Provide for the health and safety of ASM workers and their families; • Strengthen, monitor and enforce laws on child labour; • Strengthen the role and security of women in ASM; • Promote the role of ASM in rural development; and • Provide technical training to improve productivity and to safeguard the environment.

ASM is concerned with the integration of informal mining into formal economic and legislative sectors, while reducing environmental impacts and protecting the health and rights of children, women and men.

Following the conception of the IGF, a subset of interrelated sustainable development issues was making itself known globally: the water, energy and food (WEF) security nexus. The Global Risk Report of the World Economic Forum featured WEF security and climate change as recurring themes among the top five global risks in terms of impact and likelihood. In one report it was noted that "any strategy that focuses on one part of the water-energy-food nexus without considering its interconnections risks serious unintended consequences" (World Economic Forum, 2011). And in a special issue of *Water International* on Sustainability in the Water-Energy-Food Nexus, Bhaduri, Ringler, Dombrowski, Mohtar, and Scheumann (2015) described how the interplay among the security of water, energy and food is being exacerbated by growing natural resource scarcity and climate change, and how this suggests that a nexus approach to policy making would reduce costs and increase benefits compared to independent approaches.

To help operationalize the MPF in the context of WEF security in light of mining development at local and regional levels, the International Institute for Sustainable Development created the Water, Energy and Food Security Analysis Tool for Mining (WEFsat-Mining) and piloted it in Suriname, an IGF member country on the northern coast of South America. WEFsat-Mining was designed to assist mine operators, community organizations and policy makers to better understand the influence of mining on community and regional WEF security.

Box 1. Mining in Suriname.

The mining sector is by far the largest contributor to Suriname's economy, not only in terms of its contribution to government revenues, but also in relation to the country's overall economy. Additional benefits include net export earnings, infrastructure development, employment and spin-off economic activities. The resource sector focuses on oil, gold, bauxite and (of less importance) construction materials. The export value of mining commodities increased from USD 1.2 billion to USD 2.5 billion from 2007 to 2013, and represents almost 40 per cent of Suriname's GDP. Employment growth in recent years has been driven by the private sector, mainly mining. Worth mentioning is that between 20,000 and 30,000 persons are active in ASM (both formal and informal). (Source: Gemerts and Grassalco, in Roy, Bizikova, Borden, Swanson, & Huppé, 2015b)

The Water, Energy and Food Security Analysis Tool for Mining

The analysis framework used in WEFsat-Mining enables a place-based analysis of four main components: availability, access, supporting resources and supporting policies, each in the context of a region's WEF supplies (Bizikova, Roy, Swanson, Venema, & McCandless, 2013). Specifically, the activities for these four components include:

(1) Assessing how WEF are made *available* at the local level to households and communities (Table 2). This requires consideration of five aspects: (a) sources and production (e.g. surface and groundwater, sources of energy and food production); (b) water treatment, energy conversion and food processing; (c)

Table 2. Framework for assessing water, energy, and food security (Roy, Bizikova, Huppé, Borden, & Swanson, 2015a, based on Bizikova et al., 2013).

Security category	Security components to be assessed for water, energy and food sources
Availability	**Uses** **Processing** **Storage** **Distribution** **Markets**
Access	**Purchasing power** (livelihood income, remittances, credit) **Aid** (direct provision, safety nets, subsidies) **Self-production** (water wells, off-grid power, individual/community gardens) **Barter**
Supporting infrastructure	**Built infrastructure** (transportation, communication, waste removal)
	Natural infrastructure (ecosystem services such as erosion control, storm protection, water purification, biological control, air quality maintenance, pollination)
Supporting institutions and policies	**Institutions** (utility boards, user associations and resource co-ops, education and training, safety oversight, law enforcement and security)
	Policies and plans (resource use, climate change adaptation, disaster recovery, risk management, R&D and innovation)

storage of WEF supplies; (d) modes of distribution of WEF supplies; and (e) markets (both formal and informal) for WEF.

(2) Identifying how households (and communities of households) gain *access* to WEF (Table 2). Is it mostly through their own purchasing power (i.e. earned income, remittances from family members in other countries, credit), as is typical in higher-income households and countries? Or is access gained through a combination of purchasing power, aid, self-production and barter, as it often is in lower-income households and countries? Results from this component are primary pathways for gaining access to WEF.

(3) Identifying the types of *supporting infrastructure* relied on to ensure the access and availability of WEF. Supporting infrastructure has two types: built infrastructure, including communication, transportation and waste/sanitation systems; and natural infrastructure, including ecosystem goods and services such as erosion control, storm protection, water purification, biological control, soil health maintenance and pollination.

(4) Characterizing the *supporting institutions and policies* that support the natural and built infrastructure needed to ensure access and availability of WEF sources in a community and region. This component is further broken down into two categories: supporting institutions, including utility boards, user associations and resource co-ops, education and training, safety oversight, and law enforcement and security; and supporting policies and plans relating to resource use, climate change adaptation, disaster recovery and risk management, and R&D and innovation.

This analytical framework attempts to unpack the components of a complex WEF nexus at the local level and allows those involved in development (in agriculture, mining or other sectors) to identify the most effective investments and interventions to ensure minimizing negative impacts and maximizing positive ones. This, in turn, is expected to enable Suriname and other IGF member countries to mainstream social,

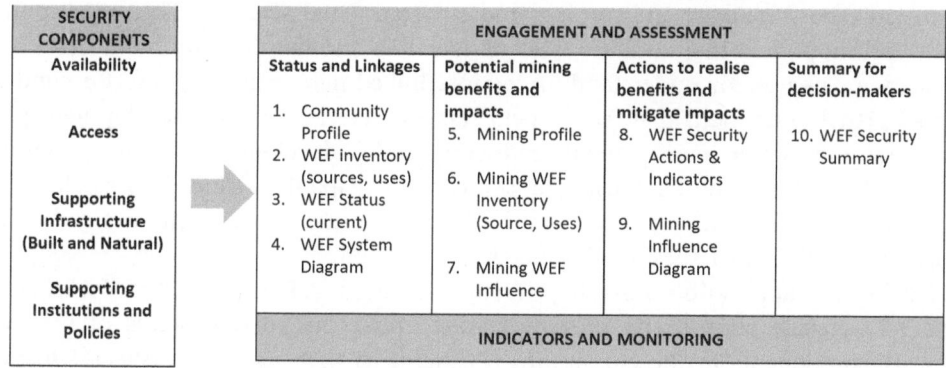

Figure 1. Architecture of the Water, Energy and Food Security Analysis Tool for Mining – numbers indicate tool worksheets (Roy et al., 2015a).

economic and environmental sustainability through the MPF into national and regional decisions related to mining.

Based on the WEF security categories, as shown in Table 2, this Microsoft Excel–based analytical tool enables community and mining stakeholders to obtain an integrated view of the potential benefits and impacts of mining operations on WEF security. The WEF security framework components are assessed through 10 Excel worksheets (Figure 1) designed to engage stakeholders in an assessment of: (1) the current status (and linkages) of the availability of and access to WEF and the array of infrastructure (built and natural) and policies that support their use; (2) the potential benefits and impacts of mining on these WEF security components; and (3) the actions necessary to realize potential benefits and mitigate impacts on WEF security components.

Essentially, the assessment tool first creates a sort of baseline for WEF security components for communities and then accounts for the impacts of mining processing (i.e. mineral extraction, mineral processing, mine waste, economic activity and general facilities) during the operation and closure phases of mine development. WEFsat-Mining also helps identify relevant indicators that can be used to track the status and trends of WEF security and the potential mining benefits and impacts, along with progress towards key actions.

Development and application of WEFsat-Mining in Suriname

Development of WEFsat-Mining in Suriname proceeded in two distinct phases. The first phase was designed to identify key local issues and to validate overlapping concerns for WEF security in the region of the state mining company's development in Maripaston, in central Suriname. Operationally, this involved convening a first workshop in Suriname, in partnership with Grassalco, the state mining company, in July 2014. These inputs informed the design and initial development of a draft version of the tool. A second phase involved refining the draft version and testing it using inputs and data obtained from the literature and through expert input in Suriname. A second workshop was convened in partnership with Grassalco and two faculties at Anton De

Kom University: geology and mining; and environmental sciences. This pilot application workshop, held in June 2015, was designed to test the draft tool and build local capacity for its use, for understanding the interlinked nature of WEF and for conducting an actual assessment that would help stakeholders involved in development planning, mining, water management and community development explore mutually beneficial solutions in the Suriname context. Participants included Grassalco staff, university professors and students, national public works department staff and NGOs working on women's issues, food security.

During the application workshop, facilitators used WEFsat-Mining Worksheets 2 and 3 to engage participants in a discussion about the current status of the WEF security components in the community. Participants then used Worksheet 4 to draw a systems map to illustrate baseline WEF usage in the community and identify the key interconnections between WEF systems (Figure 2).

Worksheet 5 was used collaboratively to define the likely profile of a proposed mine. With this profile, participants used Worksheet 6 to create an inventory of any new WEF

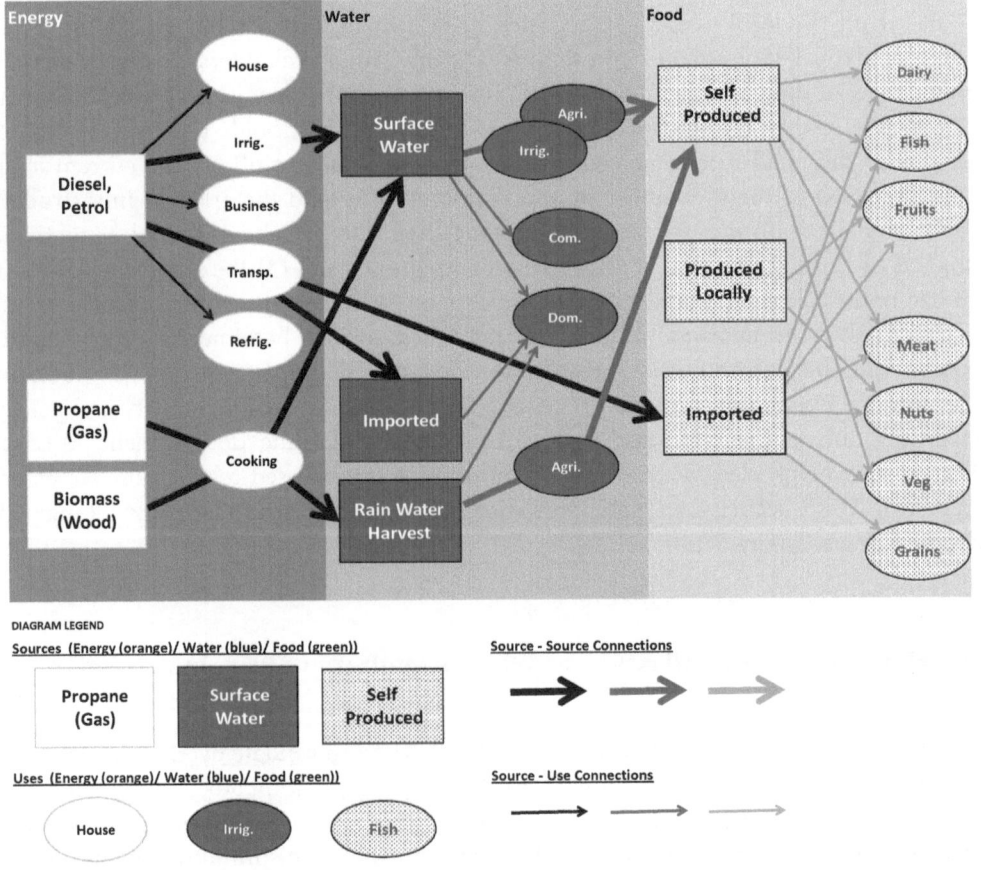

Figure 2. Water, energy and food systems map characterizing rural communities in Suriname, as generated by participants using the Water, Energy and Food Security Analysis Tool for Mining (Roy et al., 2015b).

sources introduced by mining operations (e.g. introduction of bottled water due to a new road). This helped inform a discussion to identify the potential benefits and impacts of mining processes during both the operation and closure phases on the WEF security framework components, which were recorded in Worksheet 7.

Worksheet 8 was used to summarize the benefits and impacts identified from the mining processes for each phase of mining. In addition, the indicator lists in Worksheet 8 provided participants with options for relevant state, pressure and action indicators (either chosen from a drop-down menu or self-identified) for monitoring changes from the benefits and impacts.[2] Using these findings, discussions were aimed at developing recommendations for potential actions that could reduce impacts and, where possible, improve conditions in the communities. Actions critical for improving multiple WEF security components were given priority. Though time and data gaps did not permit the participants to complete the entire analysis, examples of changes to the system were illustrated in Worksheet 9. This analysis, informed by experts through the course of the research and collaborative workshops, could be used to inform the summary findings in Worksheet 10.

All along, an attempt was made to find ways to enhance interlinked benefits and mitigate foundational challenges with impacts on WEF security. For example, improving access and sustainability of energy systems in Maripaston was seen to be clearly linked to the ability to access and use clean water and prepare food. Also, improved transportation systems are a critical intervention for all three types of security in a variety of ways. Workshop participants discussed how the purchasing power of community members could decrease after mine closure and how this would negatively impact the ability to obtain drinking water from surface water sources due to a decreased ability to buy fuel for water pumps. Potential actions to mitigate such impacts could be to introduce a surface water treatment plant and more water-efficient agriculture practices. Potential indicators to track the status of purchasing power were identified, including average annual income and the employment-to-population ratio.

Reflections on the development and application of WEFsat-Mining in Suriname

In Suriname, where mining plays a significant role in the national economy and where its impacts and implications are not well known, the WEF security lens, and methodologies and capacities to assess its interlinked systems, are of significant relevance. While a comprehensive quantitative assessment was not feasible due to limitations of data and project timeline, a generic case study using information gleaned from expert interviews and two workshops provided an understanding of the links between regional security and mining development. Further efforts to quantify these elements would be a worthwhile investment for decision makers and planners in the country. Data gaps in clearly articulating security components remain the biggest challenge in Suriname. However, with inputs from experts in different component sectors and unbiased professionals, this assessment could be completed in a few additional months. A fuller, more quantitative analysis would require more data and more investment in monitoring systems, such as the hydrologic information systems recommended as part of the research (Borden & Roy, 2015).

One of the key benefits of the WEFsat-Mining development and application work-shops, based on feedback received, was exposure to integrated thinking. Participants appreciated not just the substantive knowledge gained from explanations of WEF systems (and all their aspects) and coordinated decision making, but also learning through the participatory nature of the workshop. These workshops encouraged participants to share their perspectives with stakeholders of differing viewpoints and priorities and foster dialogue to understand each other's views. Many workshop participants found WEFsat-Mining and the processes involved very complementary to their understanding of an environmental and social impact assessment and felt that they could see themselves using WEFsat-Mining to fulfil such an assessment or conduct related analyses.

Conclusions

The MPF represents broad guidance in managing the impacts and benefits of mining for national systems – including social, economic and environmental aspects. In implementing this guidance in developed and developing countries, sub-national assessments such as the one demonstrated by the use of the WEFsat-Mining in Suriname provide a practical methodology for addressing key components of human security, as represented by WEF, as highlighted by the World Economic Forum and others.

WEFsat-Mining provides a useful and practical methodology for understanding, assessing and managing WEF security in the context of mining operations. Such efforts to assess WEF security, undertaken by governments in partnership with mining companies and other stakeholders, provide a means for realizing the various components of IGF's MPF with respect to the optimal conversion of natural capital into human capital, the management of the natural resource base within ecosystems, and continuous planning for the post-mining transition.

Mining companies, together with governments, communities, NGOs and others, play a critical role in ensuring that regional WEF security is maintained or enhanced. They have the opportunity to enable regional economic development (including skills development and local procurement), to take the lead on watershed management and water monitoring, to implement and showcase WEF security considerations as part of their corporate social responsibility, and to enhance their overall social and environmental licence to operate.

Notes

1. The focus on governments differentiates the IGF from the International Council on Mining and Metals, which also works to improve the contribution of the mining sector to sustainable development, but does so through mining companies and associations.
2. For more information on WEF security indicators, see Huppé, Bizikova, Roy, Swanson, and Borden (2015).

Funding

This work was supported by Foreign Affairs and International Trade Canada [7059907] and the Government of Canada [103879-001].

References

Bhaduri, A., Ringler, C., Dombrowski, I., Mohtar, R., & Scheumann, W. (2015). Sustainability in the water–energy–food nexus. *Water International*, *40*(5–6), 723–732. doi:10.1080/02508060.2015.1096110

Bizikova, L., Roy, D., Swanson, D., Venema, H. D., & McCandless, M. (2013). *The Water-energy-food security nexus: Towards a practical planning and decision-support framework for landscape investment and risk management*. IISD. Retrieved from http://www.iisd.org/pdf/2013/wef_nexus_2013.pdf.

Borden, C., & Roy, D. (2015). Water quality monitoring system design. Retrieved from http://www.iisd.org/sites/default/files/publications/water-quality-monitoring-system-design.pdf

Huppé, G. A., Bizikova, L., Roy, D., Swanson, D., & Borden, C. (2015). *Water-energy-food Resource Book for Mining*. IISD. Retrieved from http://www.iisd.org/sites/default/publications/water-energy-food-resource-book-mining.pdf.

IGF. (2013). A mining policy framework. Intergovernmental Forum on Mining, Minerals, Metals and Sustainable Development. Retrieved from http://www.globaldialogue.info/MPFOct2013.pdf.

Roy, D., Bizikova, L., Borden, C., Swanson, D., & Huppé, G. A. (2015b). *Water-energy-food security and mining in Suriname: A project overview*. IISD. Retrieved from http://www.iisd.org/sites/default/files/publications/water-energy-food-security-mining-suriname-project-overview.pdf.

Roy, D., Bizikova, L., Huppé, G. A., Borden, C., & Swanson, D. (2015a). *WEFsat-Mining tool user guidance manual*. IISD. Retrieved from http://delta.iisd.org/sites/default/files/publications/WEFsat-mining-tool-user-guidance-manual.pdf.

United Nations. (2002). *Plan of implementation of the world summit on sustainable development. Johannesburg, South Africa*. New York: United Nations. Retrieved from http://www.un.org/esa/sustdev/documents/WSSD_POI_PD/English/WSSD_PlanImpl.pdf

World Economic Forum. (2011). *Global risks 2011* (6th ed.). Cologne: World Economic Forum. Retrieved from https://www.weforum.org/reports/global-risks-report-2011/

Index

Note: **Boldface** page numbers refer to figures and tables, page numbers followed by "n" denote endnotes

Aarhus Convention 55, 63
acid mine drainage (AMD) 7, 17; definition 48; impact of 49; in South Africa *see* Olifants/ Limpopo River, acid mine drainage pollution
Acta de Combayo 166, 169–70
Agricultural Research Center (ARC) 183, 184
AMD *see* acid mine drainage
American Smelting and Refinery Company 127
ANA *see* Autoridad Nacional de Aguas
Ancash Region: *campesino* communities in 145, **146**; integrated water resources management framework in *see* integrated water resources management; mine activity in 146, **147**, 151–2
Ankobra River 74
Aquifer Interference Policy 36
aquifers 105; water tables of **110**
ARC *see* Agricultural Research Center
Asociación Urpichallay 149–50, 157
Astronomy Geographic Advantage Act (2007) 204
ATDRC (Cajamarca Water Authority) 170–1
Australia: mine site water-reporting practices *see* Hunter Valley coalfield; National Plan for Water Security 28; National Water Commission 42–3; 'Troubled Waters' report 30
Autoridad Nacional de Aguas (ANA) 148, 157
available water determinations (AWDs) 34–5

BOF *see* Broad Opposition Front
Bogyiri River 74
Broad Opposition Front (BOF) 124, 128, 134, 136, 138, 139

Cajamarca, water flows and distribution in 4, 103, 104, 118–20; hydrological data collection 107–9; Mashcon catchment *see* Mashcon catchment; physical alterations 104–5; socio-political alterations 105–6; Yanacocha project in 106–7

Callejon stream 112
Calueque hydroelectric project 10
campesino communities: in Ancash Region 145, **146**; integrated water resources management in *see* integrated water resources management
CARE International 150
Centre for Environmental Rights (CER) 63–4, 88
Cerro de San Pedro 4; *see also* Minera San Xavier; access to land and water 133–4; conflict in 127–8; corporate social responsibility programme 137–8; decision-making authority 135–7; echelons of rights 132–3, 139–40; land rights 134–5; landscapes and waterscapes in 128–9; Mexico's neoliberal path 126–7; natural resource management conflict 131–3; water availability in 129–31; water rights 135; ways forward 138–9
CHRAJ *see* Commission for Human Rights and Administrative Justice
coalbed methane 92
Combayo, Cajamarca 163–4; socio-environmental conflicts in 164–7, **166**; Yanacocha mining operations *see* Yanacocha gold mining company
Commission for Human Rights and Administrative Justice (CHRAJ) 77–8
commission for the monitoring of irrigation canals (COMOCA) 116
Community Mines Consultative Committee 79
Community Water and Sanitation Agency (CWSA) 74, 78, 80
COMOCA *see* commission for the monitoring of irrigation canals
Constitution of the Republic of South Africa (1996) 88
CRRC *see* Cumulative Rehabilitation Requirement Curve
Cruz, Carlos Santa 165
Cumulative Rehabilitation Requirement Curve (CRRC) 18, **18**, 19, **20**
CWSA *see* Community Water and Sanitation Agency

DAFF *see* Departments of Agriculture, Forestry and Fisheries
DCC *see* Development Cost Curve
Del Castillo, Jorge 165–6
Department of Social Development (DSD) 88
Department of Water Affairs (DWA) 92, 97, 181, 197, 200, 204, 205
Department of Water and Sanitation 97
Departments of Agriculture, Forestry and Fisheries (DAFF) 197
Departments of Mineral Resources (DMR) 197, 205
Development Cost Curve (DCC) **12**, 13
District Assembly 80, 81
DMR *see* Departments of Mineral Resources
Drilling License and Groundwater Development Regulations (2006) 79
DSD *see* Department of Social Development
Dumasi community 81
DWA *see* Department of Water Affairs

echelons of rights analysis (ERA) framework 132–3, 139–40
EIAs *see* environmental impact assessments
ejidatarios 133, 134, 136, 140
ejido system 126, 140
Electric Power Research Institute (EPRI) 92
Encajon stream 112
environmental impact assessments (EIAs) 167, 168
Environmental Protection Agency Act (1994) 80
Environment Protection and Biodiversity Conservation Act (EPBCA) 35–6
EPRI *see* Electric Power Research Institute
ERA framework *see* echelons of rights analysis (ERA) framework
Espoo Convention 55
externalization-of-costs model **12**, 13

Fanie Botha Accord 11
Fracturing Responsibility and Awareness Act (2009) 199

GEMI *see* Global Environment Management Initiative
Gender and Water Resources Management Strategy 80
Ghana Water Company Ltd. 74
Global Environment Management Initiative (GEMI) 70, 80
Global Reporting Initiative (GRI) **30**, 30–1, 36–7
Global Risk Report 211
Global Water Partnership (GWP) 147
Golden Star's Environmental Management Plan 73

goldmining industry in South Africa 7–8; constraints on water policy implementation 15–17; elements of water policy 11; externalization-of-costs model **12**, 13; policy paradigms, phases of **9**, 9–13, 21; survey of mining jurisdictions 15, **16**; unintended consequences 17–22, **18**, **20**; Witwatersrand Goldfields production, statistical analysis of **14**, 14–15
GRI *see* Global Reporting Initiative
Grootdraai Dam catchment 4, 179–81, **180**; future of 189, **191**, 191–2; land cover 183–4, 186–7, **187**, **188**; model equations 189, **190**; partial least squares analysis 179, 184–5, **188**; rainfall, evaporation, water flow 181; root mean square error 185; water quality 181–3, **182**, **183**, 186, **186**
Group for Capacity Building and Intervention towards Sustainable Development (GRUFIDES) 112
GWP *see* Global Water Partnership

Huascarán National Park 150
Huascarán Working Group 150
Hubbert Theory 14
Hunter Valley coalfield 3, 29; Aquifer Interference Policy 36; Environment Protection and Biodiversity Conservation Act 35–6; future governance of mine water 42–3; Global Reporting Initiative **30**, 30–1, 36–7; Hunter Valley salt-trading scheme 39–40; low permeability barrier 42; mine site information 33, **34**; mine water issues, analysis of 31–3, **32**; National Water Initiative framework 29; water access licences 34, **35**; water-accounting framework **30**, 30–1, **31**, 36–8; water-sharing plans 34, 38–9; watershed management **32**, 33; water take and use, indicators of **40**; water-use productivity curve 40–1, **41**, 43
Hunter Valley salt-trading scheme (HVSTS) 39–40
hydraulic fracturing 197–8, **198**

ILC *see* International Law Commission
Independent Expert Scientific Committee 35
integrated water resources management (IWRM) 5; in Ancash Region 148–9; Asociación Urpichallay 149–50; catchment scale 152; conceptual model of 151, **151**; definition of 147–8; economic sustainability 153; financial capacity 155–6; governmental decisions, Ancash of 154–5; holistic management framework 152–3; human capacity 157; institutional capacity 156; opportunities for 157–8; overview of 146–7; as 'paradigm shift' 147, 148; partnerships and local participation 153; technical capacity 156–7; Universidad

Nacional Santiago Antúnez de Mayolo
149–50; water-related issues 153, **154**
Intergovernmental Forum on Mining, Minerals,
Metals and Sustainable Development (IGF)
209, 216n1; Mining Policy Framework
209–11, **210**
International Institute for Sustainable
Development 209
International Law Commission (ILC): Draft
Articles on International Watercourses 56;
Draft Articles on Prevention of Transboundary
Harm from Hazardous Activities 55
International Mining for Development
website 73
IWRM *see* integrated water resources
management

Johannesburg 7, 8; mine residue areas 14, 17
Junta de Usuarios del Río Chonta
(JURCH) 173n4

Kumasi ventilated-improved pit (KVIP) 75

Loredo, Oscár 129
loss of productivity 91
low permeability barrier (LPB) 42

Mann-Kendall test 108
Mansi River 74
Marikana crisis 12, 21
Mashcon catchment 106, 120; Grande River
109–11; groundwater 109; hydrography of
111; irrigation canals in 116–18, **117**, **118**;
location of **107**; multisectoral water rights
113; MYSRL water use 114–15; rural and
urban populations 115–16; streams and
springs 111–13; water distribution and users
in 113–14
Mexico 126–7; Agrarian Law 134, 135; Mining
Law 126, 134, 135; neoliberal path 126–7
Mine Profit Curve (MPC) **18**, 19, **20**
Mineral and Petroleum Resources Development
Act of 2002 (MPRDA) 97, 200, 206
Minerals Council of Australia (MCA) 30
Minera San Xavier (MSX) 124, 128, 133,
134, 140; corporate social responsibility
programme 137–8; opportunities offered by
130–1; subsistence and opposition 130
mine residue areas (MRAs) 14, 17
Mine Revenue Curve (MRC) **18**, 19, **20**
mine site water-reporting practices, in Australia
see Hunter Valley coalfield
Mining Livelihoods Curve (MLC) **12**, 13
Mining, Minerals and Sustainable Development
(MMSD) 70
Mining Policy Framework (MPF): pillars of
209–11, **210**; in water-energy-food nexus *see*

Water, Energy and Food Security Analysis
Tool for Mining
Ministry of Water Resources Works and Housing
(2007) 80
MLC *see* Mining Livelihoods Curve
MMSD *see* Mining, Minerals and Sustainable
Development
MPC *see* Mine Profit Curve
MPF *see* Mining Policy Framework
MPRDA of 2002 *see* Mineral and Petroleum
Resources Development Act of 2002
MRAs *see* mine residue areas
MRC *see* Mine Revenue Curve
MSX *see* Minera San Xavier
MYSRL 106, 107, 109; environmental
impact assessments of 112–13; water use
and 114–15

NAFTA *see* North American Free Trade
Agreement
National Environmental Management Act of
1998 (NEMA) 62, 88
National Environmental Management: Waste Act
(NEM:WA) 62
National Framework for Compliance and
Enforcement Systems for Water Resource
Management 35
National Plan for Water Security
(Australia) 28
National Planning Commission 86
National Water Act of 1998 (NWA) 62, 97
National Water Authority *see* Autoridad Nacional
de Aguas
National Water Commission (NWC) 42–3
National Water Initiative (NWI) 29, 34
National Water Policy 80
NEMA *see* National Environmental Management
Act of 1998
New South Wales (NSW) 29, 33–7, 40
Non-Mining Livelihood Curve (NMLC) **12**,
13, 21
North American Free Trade Agreement (NAFTA)
5, 126–7, 136, 140
NSW *see* New South Wales
NWC *see* National Water Commission
NWI *see* National Water Initiative

Observatory of Mining Conflicts in
Peru 106–7
oil-from-coal technology 10
Olifants/Limpopo River, acid mine drainage
pollution 4, 50–1; Centre for Environmental
Rights, role of 63–4; environmental impacts
52; historical legacy of 51–2; transboundary
impacts 52–3; United Nations Watercourses
Convention *see* United Nations Watercourses
Convention

partial least squares (PLS) regression 179, 184–5, **188**
PED-nexus *see* population–environment–development nexus
Peru, conflict resolution strategies in *see* Yanacocha gold mining company
Peruvian Water Law 113
physical water snapshot 71, 73–4
PLS regression *see* partial least squares regression
"p-N Cycle" **14**
polluter-pays principle 15, 22
population–environment–development nexus (PED-nexus) 87–91, **89, 90**; water-related impacts in 90, 91
Prestea Gold Resources 71
Pulp Mills on the River Uruguay (Argentina v. Uruguay) 55, 58, 61

RC *see* Revenue Curve
RCC *see* Remediation Cost Curve
Remediation Cost Curve (RCC) **12**, 13
Revenue Curve (RC) **12**, 13
root mean square error (RMSE) 185

SADC Revised Protocol 54, 57
SAIRR *see* South African Institute of Race Relations
SANBI *see* South African National Biodiversity Institute
San Luis Potosí 128, 129, 135
SEAs *see* strategic environmental assessments
SEDACAJ 115–16
shale gas mining 4; activity to monitor and regulate 198–200, **199**; adoption of policies 204; Astronomy Geographic Advantage Act 204; Departments of Mineral Resources report 203; departments responsible for regulatory activities 204–5; "extremely useful tool" 202, **202**; fracking-specific monitoring protocols 202, 206; hydraulic fracturing 197–8, **198**; National Development Plan 203; overview of 195–6; self-regulation 204; South Africa's regulatory regime 200–1, **201**, 205; strategic environmental assessments 203; structured questionnaire 196–7; study objectives 196
"significant transboundary harm" 59–60
Small Communities Water and Sanitation Policy 80
social impact assessment 70
Social Water Assessment Protocol (SWAP) 5; climate conditions 74; domestic uses 75–6; formal and informal economy 76; gender 78; Golden Star's Environmental Management Plan 73; health issues 78–9; human rights landscape 77–8; indigenous people 76; interaction between stakeholders 79; key

stakeholders 79; local amenities **75**; physical water snapshot 71, 73–4; purpose of 70; social, cultural and spiritual 77; theme of **72, 82**; water supply and infrastructure 74–5
socio-environmental conflicts: in Combayo 164–7, **166**; water governance and 162–3
Solanes, Miguel 136, 137
solid waste and wastewater 94
South Africa: acid mine drainage pollution *see* Olifants/Limpopo River, acid mine drainage pollution; economy 91; environmental policy framework 87, 88; goldmining industry *see* goldmining industry in South Africa; Grootdraai Dam catchment *see* Grootdraai Dam catchment; National Environmental Management Act of 1998 88; population 86; shale gas mining in *see* shale gas mining
South African Institute of Race Relations (SAIRR) 92
South African National Biodiversity Institute (SANBI) 183, 184
South African National Water Resource Strategy 96
Standard Land Cover Classification Scheme 183
strategic environmental assessments (SEAs) 203
Strategic Regional Land Use Policy and Plans 36
Suriname: mining sector in 211; Water, Energy and Food Security Analysis Tool for Mining 213–16, **214**
sustainable development 88
SWAP *see* Social Water Assessment Protocol

Trail Smelter Arbitrations (US v Canada) 55
Tripartite Alliance 12
'Troubled Waters" report 30

Ulan underground mine 38–9
UNASAM *see* Universidad Nacional Santiago Antúnez de Mayolo
unconventional oil and gas (UOG) resource 4, 86; agricultural livelihoods and food security 92–3; energy development and demand for water 91–2; extraction and management 87; human populations 93–4, 97; policy implications 96–7; population–environment–development nexus 87–91, **89, 90**; quantity and quality, of water 93–4; scale of impacts 94–6; in South African *see* South African; water requirements for 93
UNECE *see* United Nations Economic Commission for Europe
UNGA *see* United Nations General Assembly
United Nations Economic Commission for Europe (UNECE) 55
United Nations General Assembly (UNGA) 69
United Nations Watercourses Convention (UNWC) 3; Article 32 of 49, 50, 54; claim for

liability and compensation 57, 61–2, 64–5; establishing causation 60–1; imminent threat/actual occurrence of transboundary harm 58–9; non-discrimination, principle of 55–7; "no significant harm" 53, 54; reasonable and equitable utilization 53; SADC Revised Protocol 54, 57; "significant harm," notion of 59–60

United Sates: disclosure of chemicals 199–200; Fracturing Responsibility and Awareness Act (2009) 199; groundwater contamination in 200; Safe Drinking Water Act 199

Universidad Nacional Santiago Antúnez de Mayolo (UNASAM) 149, 150

UNWC *see* United Nations Watercourses Convention

UOG resource *See* unconventional oil and gas resource

Valle de San Luis Potosí aquifer 129

WAF *see* water-accounting framework

water-accounting framework (WAF) **30**, 30–1, **31**, 36–8

Water Act (1956) 10

water and sanitation management (WATSAN) teams 78, 80

water availability 87

Water, Energy and Food Security Analysis Tool for Mining (WEFsat-Mining) 211, 216; analytical framework and components of 211–12, **212**; architecture of 213, **213**; in Suriname 213–16, **214**

water–energy–food nexus 5

water footprint 71

water governance: definition of 162; and socio-environmental conflicts 162–3

water management 70

Water Resources Commission Act (1996) 79

water security, definition of 28

Water Services Act (WSA) 62

water-sharing plans (WSPs) 34, 38–9

water tables of aquifers 109, **110**

WATSAN teams *see* water and sanitation management teams

WEFsat-Mining *see* Water, Energy and Food Security Analysis Tool for Mining

Witwatersrand Goldfields, statistical analysis of production **14**, 14–15

Yanacocha gold mining company 4, 103, 118–19, 164; Acta de Combayo 166, 169–70; ATDRC, role of 170–1; authorization from Regional Agriculture Authority 164–5; Carachugo expansion project 167; conflict resolution strategies 171–3; environmental impact assessments 167, 168; environmental impacts of 103–4; farmers protest against 164–7, **166**; irrigation-water users in 106; operating according to law 167–9; scientific/technical impact assessments 169–71

zona de veda 129, 141n1